Tutorial for Soldering Beginners

Implemented by SparkTip

Tangible and Accessible Solution for Electronics Edtech

EIM Technology Ltd.

Ordering Information:

Quantity sales of the published workpiece. Special discounts are available on quantity purchases by schools, academic institutions, corporations, associations, and others. For details, contact the publisher at the information below.

EIM Technology

180 - 6660 Graybar Rd,

Richmond, BC, V6W 1H9 Canada

service@eimtechnology.com

www.eimtechnology.com

EIM Technology is a registered trademark of EVO-IN-MOTION Technology Ltd.

Table of Contents

I - Introduction 1

    Overview 2

    THT and SMT Soldering 3

II - Get Familiar with Tools 5

    Soldering Iron 6

    Solder Wire 7

    Solder Fume Extractor 9

    Iron Tip Cleaner 10

    Other Commonly Used Accessories 12

III - Solder Two Pieces of Wires 17

    Wire Basics 18

    Simple Soldering 20

    Applying Flux 22

    Finishing 24

IV - Solder Through-hole Components 27

    Through-hole Component Basics 28

    Soldering on Perfboard 28

    Solder a Simple Circuit with Perfboard 33

    Desoldering THT Components 36

    Project - LED Chaser 41

    Project - Metal Detector 44

    Project - Luck Wheel 47

V - Surface Mount Technology Basics 51

    Surface Mount Technology Basics 52

    Prepare for SMT Soldering 53

    Soldering SMT with Two Terminals 55

Soldering More Complicated Components 58

Practice Board 64

Project - Fidget Spinner 67

Apx 1 - Common Mistakes to Avoid 71

Solder Bridging 72

Excessive Solder 73

Balling 73

Cold Joint 74

Overheated Joint 74

Insufficient Wetting 75

Lifted Pads 75

Solder Starved 76

Tombstoning 77

Solder Skips 77

Apx 2 - Components and Symbols 79

SMD Dimensions 80

Common Components & Footprints 81

A multilayer PCB Stackup 84

1 - Introduction

In electronics, soldering skills are essential for assembling and repairing electronic circuits. Whether you're a hobbyist or a professional, mastery of soldering skills is a priceless asset for circuit testing, debugging, and DIY projects.

Overview

Technically, soldering is the process of joining metal workpieces using fusible metal alloy called solder, to create strong and reliable electrical connections.

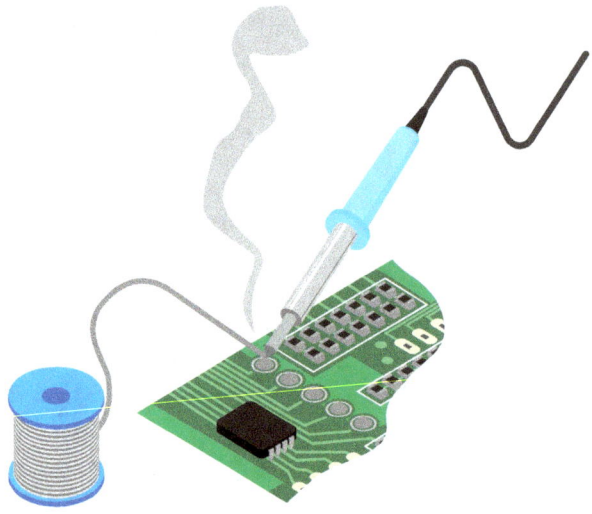

Figure 1.1: Soldering on a Printed Circuit Board

In this tutorial, we will teach fundamental soldering techniques and guide you step by step with illustrations and examples.

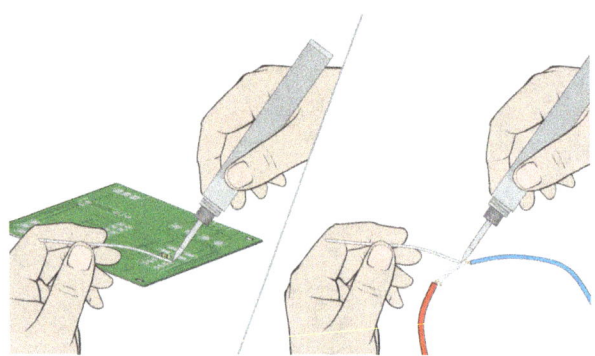

Figure 1.2: Soldering components and wires

We will also explore various soldering methods and examples of common mistakes.

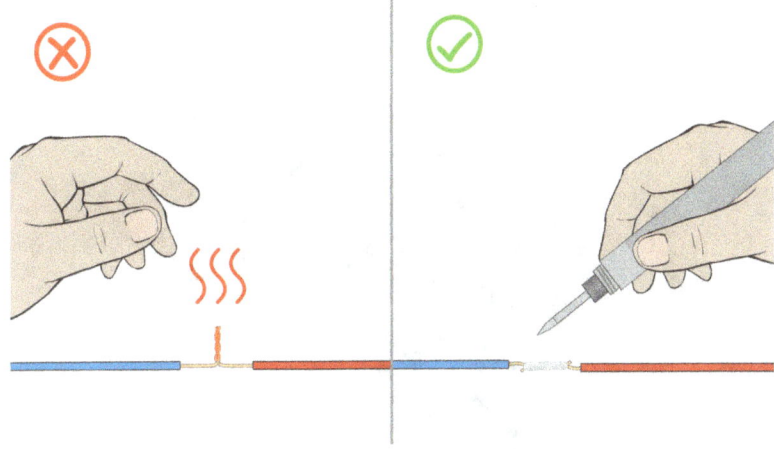

Figure 1.3: Handle parts with caution during soldering

Nevertheless, do keep in mind that soldering is a skill that highly relies on hands-on experience, so make sure you get plenty of practice before working on sophisticated projects.

THT and SMT Soldering

There are two mainstream types of soldering techniques: through-hole technology (THT) and surface-mount technology (SMT), each with its own unique characteristics and applications as listed in the Table 1.1.

TABLE 1.1

Technique	Description	Advantages	Disadvantages
Through-Hole Soldering	Inserting component leads through pre-drilled holes in a circuit board	Strong, reliable connections; handling higher power applications	Time-consuming
Surface-Mount Soldering	Soldering components directly onto the surface of a circuit board using small metal pads	More efficient and less time-consuming; allows for tiny components	Requires specialized equipment to assemble and inspect

For example, Figure 1.4 shows the circuit board inside Zoolark, a multifunctional circuit debugger by EIM Technology. This circuit board is small but integrated with a complex circuitry.

Figure 1.4: The hardware board design of Zoolark

In general, SMT is more popular due to its smaller size, higher component density, and better electrical performance, but THT also plays a role for its excellent mechanical strength and higher current capability.

You will learn both techniques in later sections of the book.

II – Get Familiar with Tools

A professional-grade set of soldering tools and materials can be extensive, however, the most essential item required to start soldering is a soldering iron.

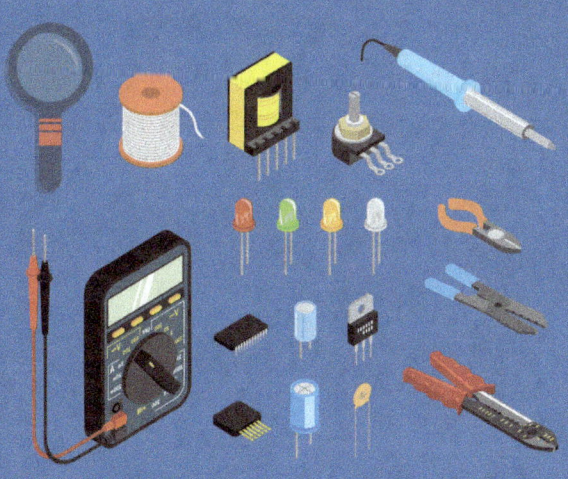

Soldering Iron

A soldering iron is a handheld tool that produces heat to join two metal workpieces together. All soldering irons come with a handle and iron tip. While operating, the tip can go up to hundreds of degrees Celsius so never touch the tip with your fingers.

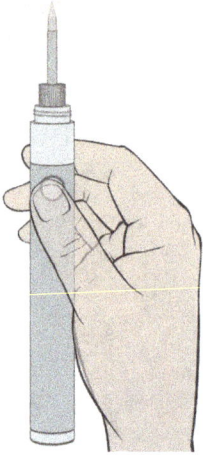

Figure 2.1: A soldering iron contains a handle and an iron tip

A full soldering iron set usually comes with a stand to place the iron when it is not in use. Note that the tip should always be pointed in the direction of a safe area.

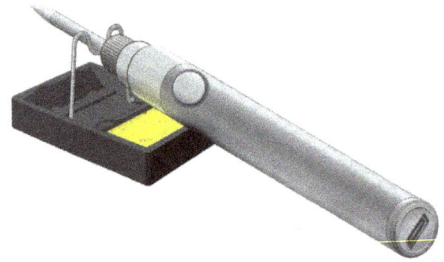

Figure 2.2: A soldering stand

Soldering tips come in various types, including knife, bent, conical, chisel and bevel, each with its own unique shape and purpose. You will see their differences in action in later practices.

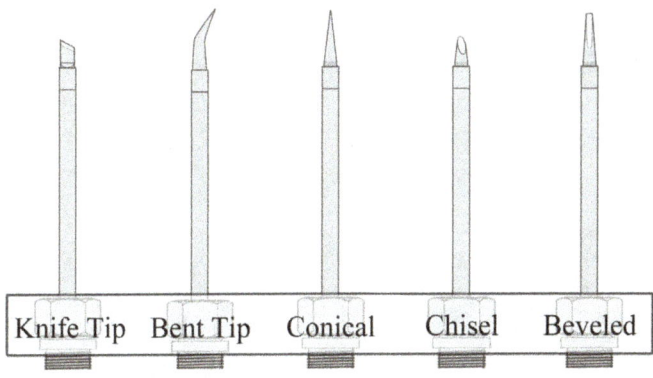

Figure 2.3: Different types of soldering iron tips

Solder Wire

You also need a solder wire to get started.

Figure 2.4: Joining wire leads to connector pins

Solder wire is a thin, coiled metal wire coated with a mixture of tin and lead that melts when heated. When the melted solder wire cools down and solidifies, it forms a strong and electrically conductive bond between the metal components being joined.

For tin-lead solder alloys, the melting temperature can range from around 180°C to 230°C (356°F to 446°F), depending on the specific ratio of tin to lead in the alloy. Lead-free solder alloys typically have higher melting temperatures, typically ranging from 220°C to 260°C (428°F to 500°F).

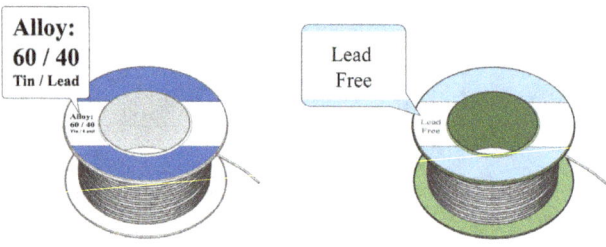

Figure 2.5: Solder wire spool

Due to the lower melting point, lead solder wires are relatively easier to work with, but lead can impose potential health and environmental risks. Be aware that even lead-free solder wire still contains harmful substances such as rosin, flux and metal oxides.

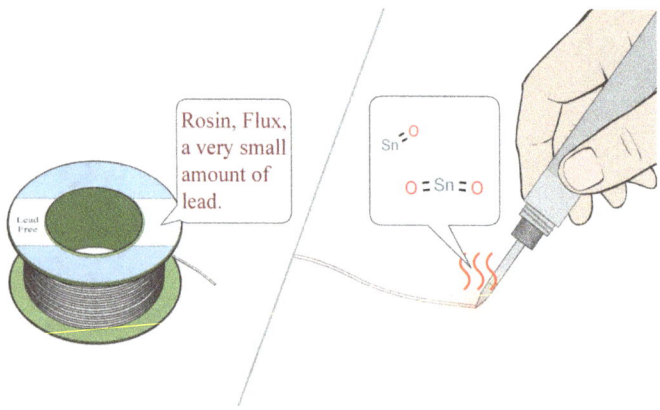

Figure 2.6: Harmful substances generates from soldering

Solder Fume Extractor

While soldering, the heat can cause the flux to release fumes that may contain harmful chemicals. Inhaling fumes over a prolonged period can lead to health problems such as respiratory issues and headaches. Therefore we always recommend using a fume extractor or fan to blow the fume away.

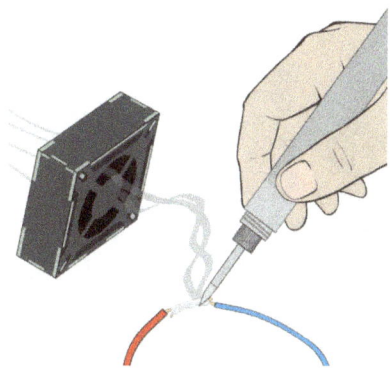

Figure 2.7: Use a fume extractor or fan to blow away the fume

If the soldering task takes a long time, you should also wear a mask and turn on the fan to direct the fumes towards a well-ventilated area or a space where no people are present.

Figure 2.8: Stay at a well ventilated area or wear a mask during soldering

Iron Tip Cleaner

Regular cleaning of the tip before and after use can also maintain its effectiveness and longevity. To clean the solder tip, you can use either brass wool or a wet sponge. A wet sponge is cheaper but needs more frequent replacement, here we prefer brass wool due to its durability and robust performance.

Figure 2.9: Brass wool and Sponge

To clean, rub the tip multiple times inside the brass wool to remove excess solder and debris. A good tip should appear silver and shiny.

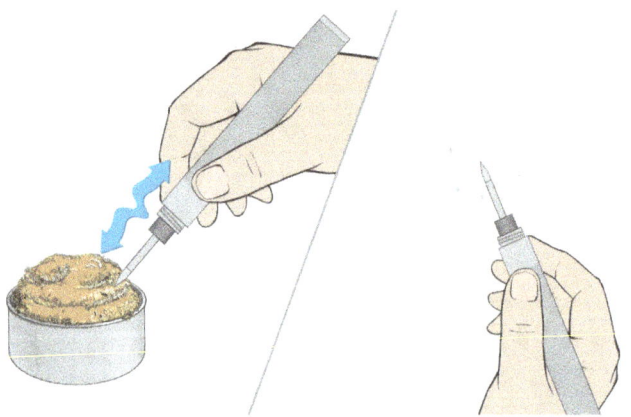

Figure 2.10: Cleaning the tip

Moisture in the air can cause oxidation of the soldering tip surface, which reduces its ability to transfer heat effectively. Therefore, after cleaning the soldering tip, it is also recommended to apply a small amount of solder to the tip to prevent direct exposure in the air.

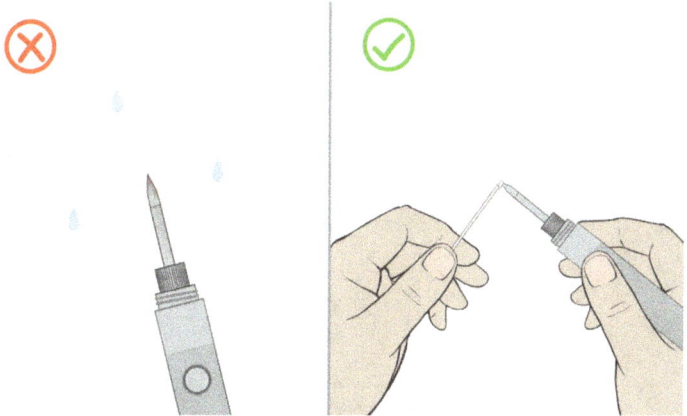

Figure 2.11: Apply some solder tin on the tip after use

When you are completely finished with soldering, make sure the soldering iron is cleaned, powered off and placed in a safe location. Consider using a protective cover or case if possible.

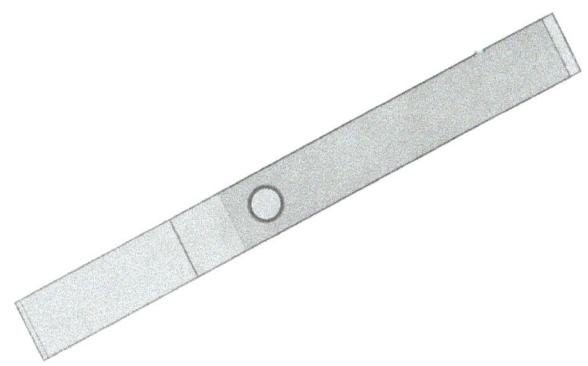

Figure 2.12: Make sure soldering iron is turned off and stored safely after use

Other Commonly Used Accessories

Aside from the necessary tools, there are additional soldering tools and accessories that aid in the soldering process. In later sections, we will discuss how these tools and accessories are used in practical soldering applications.

Flux

Flux is a chemical agent that helps the flow of molten solder and improves the quality of the solder joint. Flux comes in various forms, including liquid, paste (usually made of rosin) and flux-cored solder wire.

Figure 2.13: Rosin-based flux liquid, paste and flux core solder wire

Tweezers

Tweezers are small, handheld tools commonly used in soldering to precisely position and hold small components, wires, or solder. They are a must-have tool in SMT soldering, which you will see in later sections.

Figure 2.14: Electronic tweezers, bent and flat tips

Helping hands

A helping hand stand is another useful tool to assist soldering. It typically consists of a base with two or more adjustable arms that can be moved and locked into place to hold the item being soldered.

Figure 2.15: Helping hands station

During soldering, your hands are occupied with soldering iron and solder wire, therefore a helping hand stand can assist positioning the board and workpieces of your components.

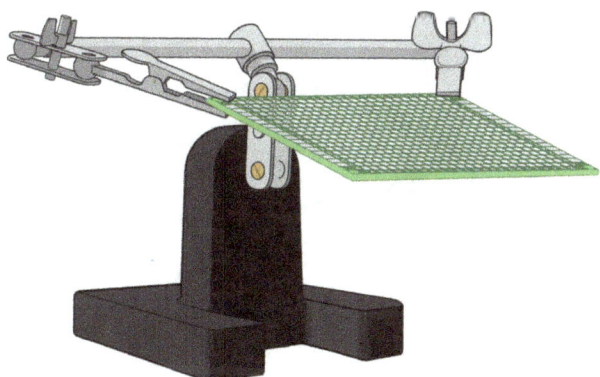

Figure 2.16: Using helping hands to hold the board to free your hands

Goggles

Goggles can effectively protect the eyes from potential hazards such as hot solder, debris, or harmful chemicals. They are an essential safety tool for anyone working with potentially hazardous materials or equipment.

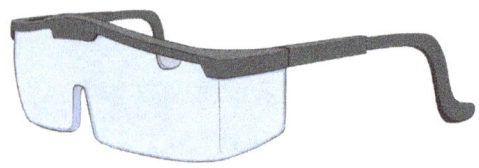

Figure 2.17: Wearing goggles during soldering is recommended

Kapton tape

Kapton tape is a high-temperature resistant tape made from a polyimide film. It may protect delicate components from heat damage, as well as to secure wires and other materials in place during the soldering process.

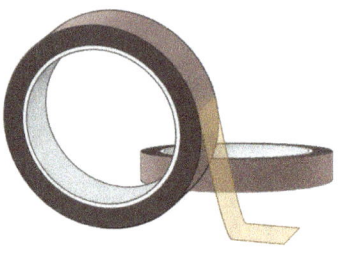

Figure 2.18: Kapton tape provides insulation and protection

Heat shrink

Heat shrink is a type of tubing made of polymer material that shrinks down to fit tightly around an object when heated. It is commonly used in soldering to protect and insulate electrical connections, as well as to reinforce weak points in wires or cables.

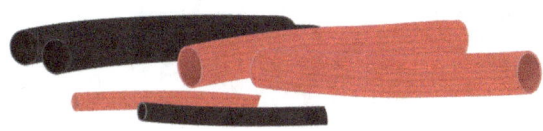

Figure 2.19: Heat shrink

Heat shrink tubing comes in different sizes and colors and is applied using a heat gun or lighter.

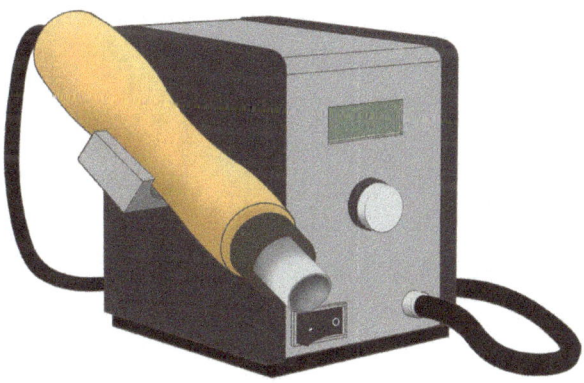

Figure 2.20: Hot air gun

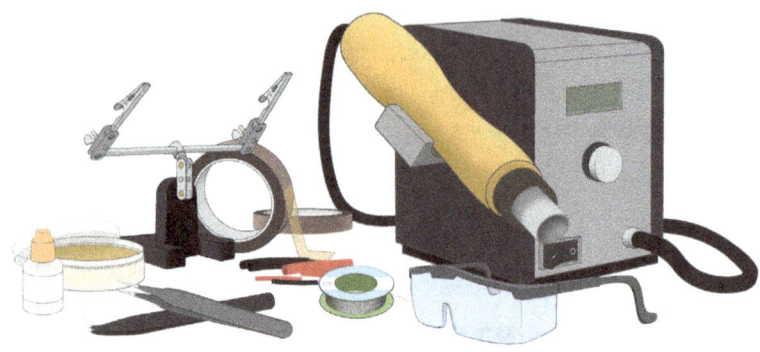

Figure 2.21: A complete set of tools often seen on a soldering workbench

III - Solder Two Pieces of Wires

Wire is a thin and flexible piece of conductor, typically copper or aluminum, that is used for conducting electrical current. Wires usually have an insulation jacket that covers the metal to prevent current from flowing to unintended places and also protect the wire from damage.

Wire Basics

In electronics applications, we may see two types of wires: stranded wires and solid wires.

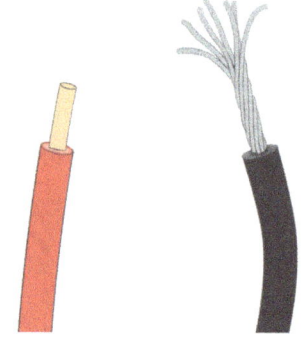

Figure 3.1: Solid wire and stranded wires

Solid wires are stiffer and hold their shape better, making them ideal for breadboard circuiting or other applications that require stiffness of the wires.

Figure 3.2: Solid wires on breadboard

Instead of a single solid core, stranded wires are made up of multiple thin metal cores twisted together, which gives more flexibility and durability. Stranded wires are ideal for applications that require frequent movement or bending.

Figure 3.3: Stranded wires contain multiple flexible thin metal cores

The current-carrying capacity of wires is related to the size of wires, which are characterized by wire gauge. A standardized system used to measure the size of wires is the American Wire Gauge, known as AWG number. A bigger AWG number indicates thinner wires.

TABLE 2.1 - AWG Conversion Table

AWG (N°)	Diam. (mm)	Area (mm^2)	AWG (N°)	Diam. (mm)	Area (mm^2)
1	7.35	42.40	16	1.290	1.310
2	6.54	33.60	17	1.150	1.040
3	5.83	26.70	18	1.024	0.823
4	5.19	21.20	19	0.912	0.653
5	4.62	16.80	20	0.812	0.519
6	4.11	13.30	21	0.723	0.412
7	3.67	10.60	22	0.644	0.325
8	3.26	8.350	23	0.573	0.259
9	2.91	6.620	24	0.511	0.205
10	2.59	5.270	25	0.455	0.163
11	2.30	4.150	26	0.405	0.128
12	2.05	3.310	27	0.361	0.102
13	1.830	2.630	28	0.321	0.080
14	1.630	2.080	29	0.286	0.065
15	1.450	1.650	30	0.255	0.050

Most breadboard wires have AWG between 22 to 24. Wires used to carry higher current tend to be thicker and therefore have lower AWG.

Simple Soldering

Joining two pieces of wire is one of the most basic soldering skills for engineers and hobbyists, you will need to apply this skill when repairing circuit boards, replacing damaged wires, connecting components and many more scenarios.

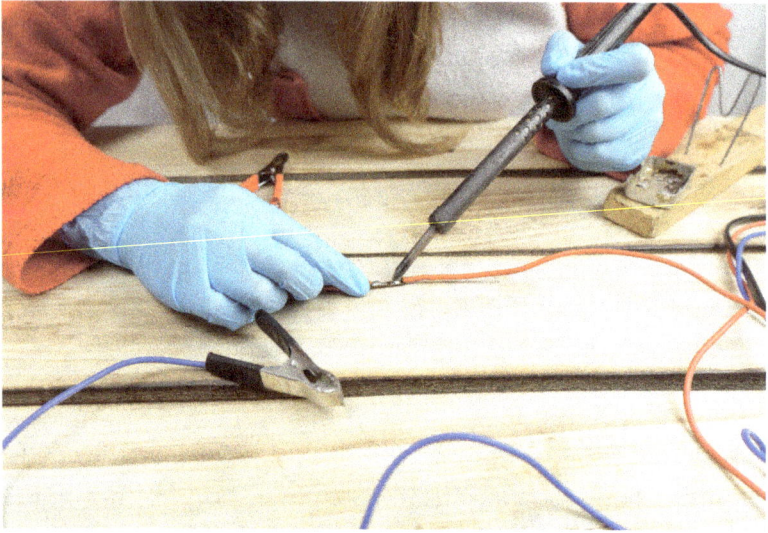

Figure 3.4: Joining two pieces of wires using solder

If the wire is fully covered by the isolating jack, use wire strippers to remove about 1/2 inch of the insulation from the ends of both wires. The wire strippers should have multiple slots for different wire gauges, make sure to use the appropriate AWG for your wire.

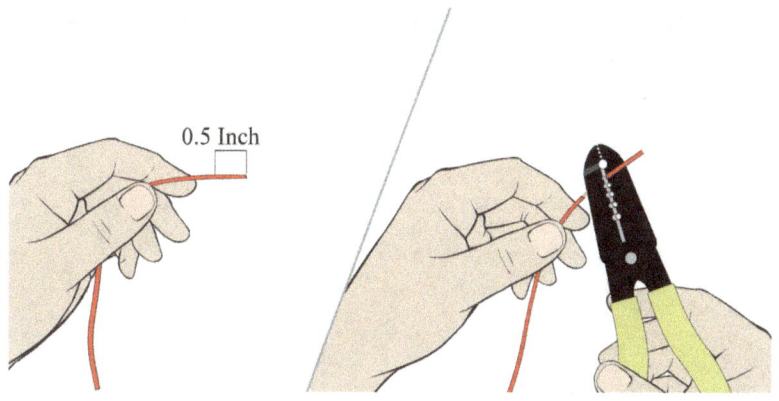

Figure 3.4: Using a wire stripper to remove a segment of insulating jack

Twist the wires together so that they form a tight mechanical connection. This step is the same for both solid wires and stranded wires.

Figure 3.5: Twisting two wires to make a tie mechanically

Turn on the soldering iron and wait until the tip heats up. Note that the entire soldering iron tip gets heated very quickly (usually less than 30 seconds) so you never want to touch the iron tip.

Figure 3.6: A working iron tip is a few hundred degrees Celsius

Meanwhile, consider securing the wire using a helping hand soldering stand to release both of your hands for soldering. Then you can apply solder to the metal and join the two wire leads electrically.

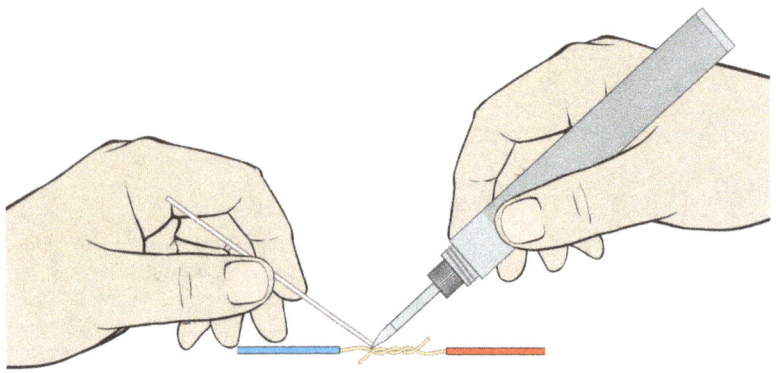

Figure 3.7: Applying solder to the mechanically tied wire leads

Applying Flux

If the solder wire is exposed to the air for a while, the surface will form an invisible but thin oxidized layer, which will result in the solder only sitting on the surface of the wire core but not really bonding to the copper.

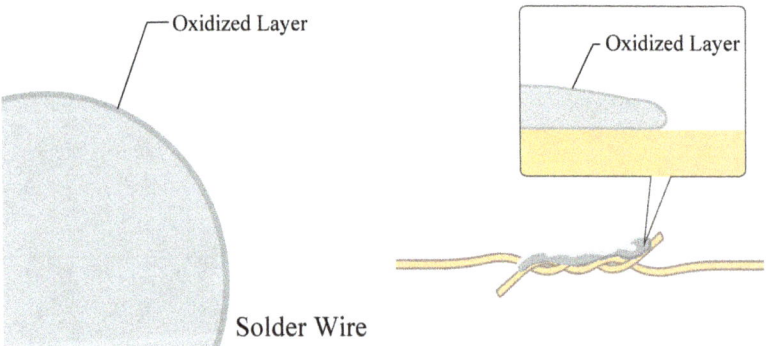

Figure 3.8: The oxidized layer prevents solder from bonding with the copper

To assist soldering, it is common to apply some solder flux, which is typically composed of rosin and is available in either a yellowish paste or liquid form. The flux helps to remove impurities and ensure a strong bond between the metal surfaces.

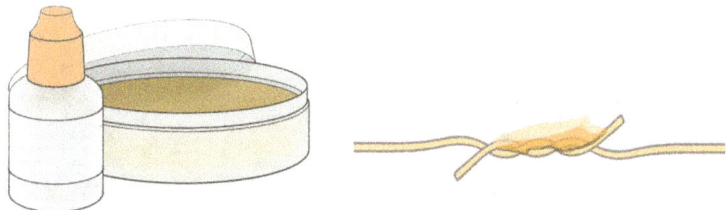

Figure 3.9: Apply some flux onto the copper to remove impurities

Once the flux is applied to the joint, repeat the soldering process again with iron and solder wire. This time the solder should be able to reach all the way.

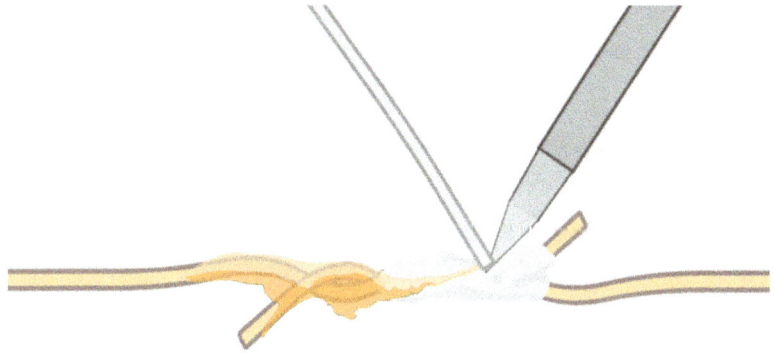

Figure 3.10: Flux provides a better thermal conduction

Let the solder cool and inspect your work. A good soldering joint should be smooth and shiny. Use a wire cutter to cut sharp leads.

Figure 3.10: Sharp tips can produce antenna effects for high frequency signals

Finishing

In most cases we want to apply an insulation jacket to the joint to prevent accidental contact and damage caused by moisture and dust. One simple way to do this is by wrapping the joint in electrical tape.

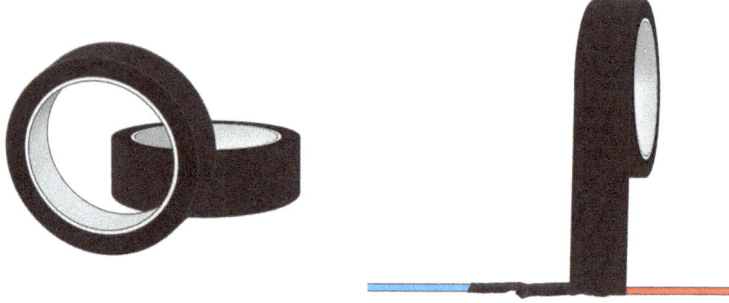

Figure 3.11: Using insulating tape (electric tape) to cover the metal joint

A more professional and reliable way is to place heat shrink tubing over the soldered joint.

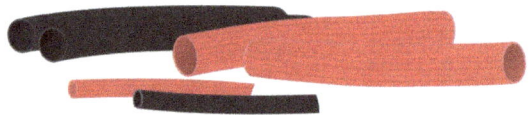

Figure 3.12: Choose heat shrink tubes of appropriate dimeters and lengths

However, you need to slide the heat shrink over one of the wires before twisting and soldering them together. This is because once the wires are joined, you cannot slide the tubing over the joint without damaging the soldered connection.

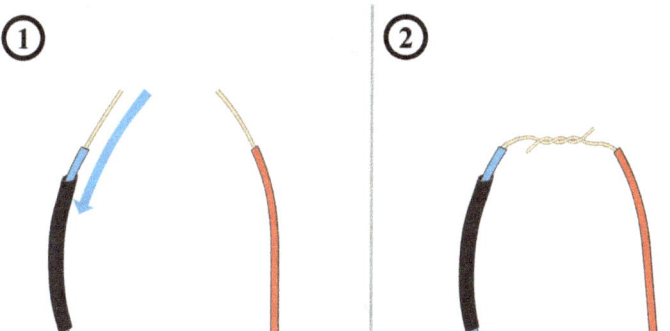

Figure 3.13: Slide the tubing on first, before tightening the two wires

Once soldering is finished, you need to ensure the heat shrink tubing is properly sized and fitted to the joint, which you can do by heating it with a hot air gun or lighter.

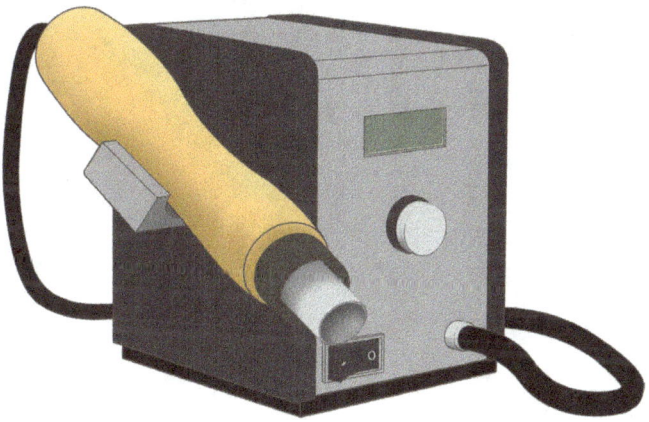

Figure 3.14: Using a hot air gun can quickly shrink a tube

If a hot air gun is not available, you can also use your soldering iron to shrink the tube.. Use the hot surfaces of the iron to heat tube evenly until it is tightly fitted to the joint. Avoid overheating the tube or the wires.

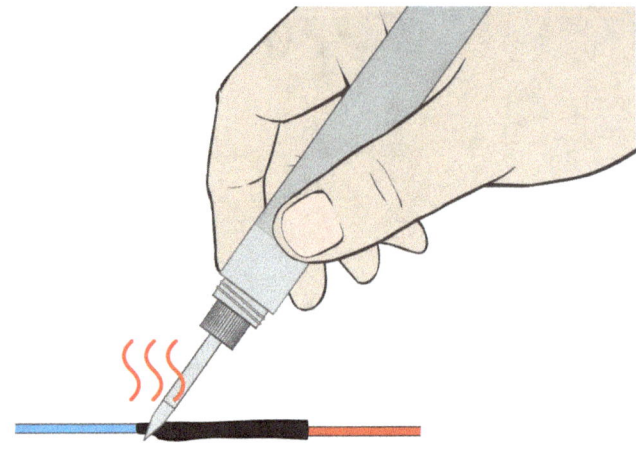

Figure 3.15: Soldering iron can also shrink the tube but takes a bit longer

Note that if you are soldering very thick wires, we recommend using a high power soldering iron and thicker solder wires.

IV - Solder Through-hole Components

Through-hole technology (THT) was widely used in the early days of electronics manufacturing before surface mount technology (SMT) became widespread. THT components were originally used in electronic equipment such as radios, TVs and computers.

Through-hole Component Basics

THT components have metal leads or pins that extend from the
component, which can connect to other component leads or wires. Some
small size THT components can also directly plug-fit into a breadboard.

Figure 4.1: Through-hole components connected onto a breadboard

THT parts are often seen in hobby electronics and technology education
because they offer effective heat dissipation and current capacity, and are
easier to manually solder or connect without specialized tools.

Soldering on Perfboard

A perforated circuit board (or perfboard) is a type of circuit board with
a grid of holes that allow components to be inserted and soldered.
Perfboards are commonly useful for testing and refining designs before
creating a final product on a printed circuit board (PCB).

Figure 4.2: An electronic perfboard

To work with a perfboard, a soldering iron and solder wire are required, and it is advisable to have a helping hand stand for assistance during the soldering process.

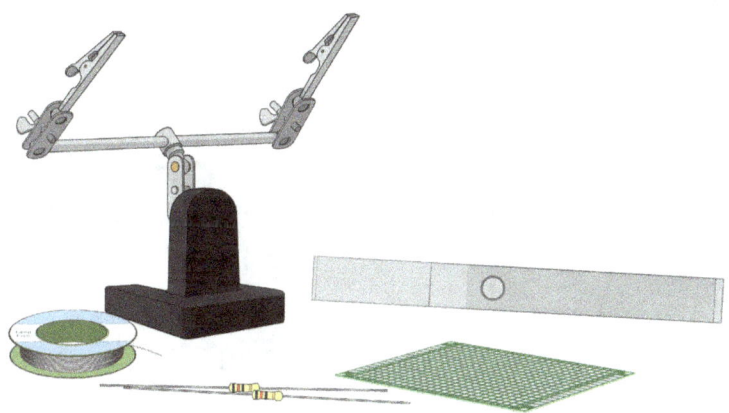

Figure 4.3: Preparation for perfboard soldering

When adding a THT part onto the perfboard, you need to bend its leads and insert the part from one side of the board.

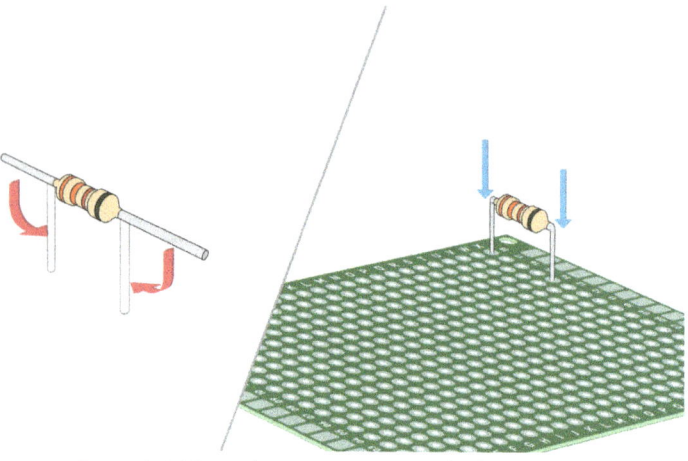

Figure 4.4: Mount the component onto the board

On the other side, bend the leads to hold the component in place.

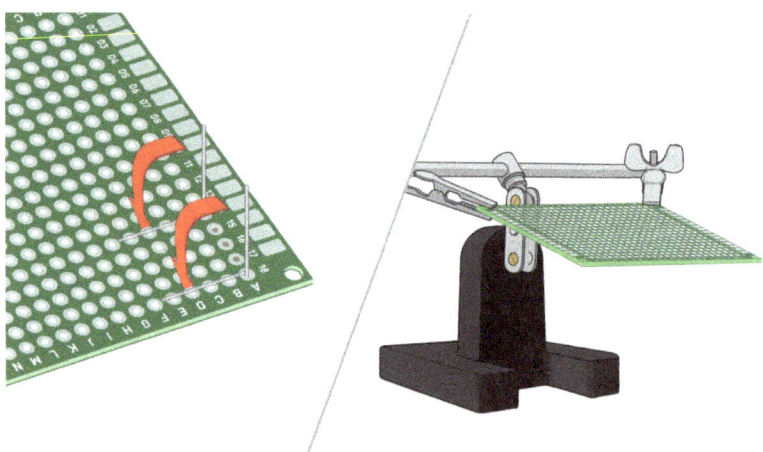

Figure 4.5: Flip over the board

Make sure the component does not fall off when the board is flipped over.

Heat up the soldering iron and solder the component lead to the tin pads on perfboard.

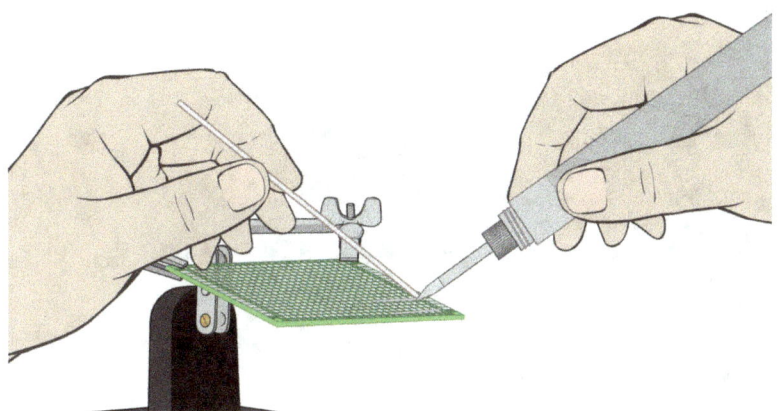

Figure 4.6: Use a sharper tip is easier

To solder a new component with an electrical connection to the existing one, bend the leads of the new component and place it in an adjacent spot. Apply heat and solder to ensure a good connection between the leads.

Figure 4.7: Bend the leads to create electrical connections

Sometimes you will need to cross over your existing paths, but you do not want to connect them. To do this, you will need to add a jumper wire. You can place the jumper wire on either side of the board but make sure the wires are insulated from others.

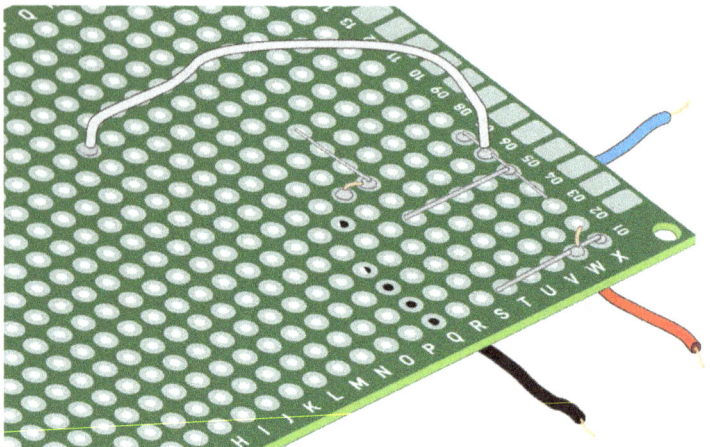

Figure 4.8: Apply a jumper wire to the board

After completing the task, wait for a moment to allow the components to cool down, or you can use a fan to speed up the cooling process.

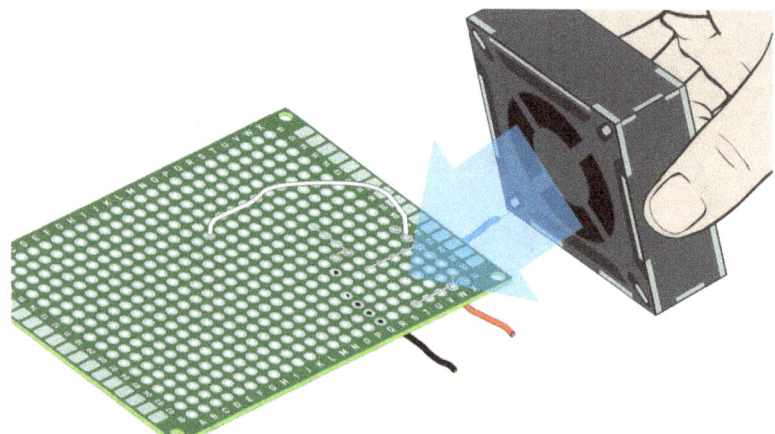

Figure 4.9: Cool the board with a fan

As we mentioned earlier, sharp wire can produce unwanted antenna effects that cause electromagnetic interference, so it is a good habit to trim excessive wire leads after soldering.

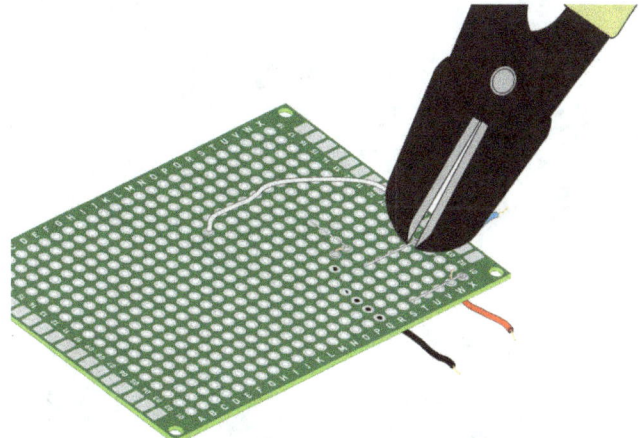

Figure 4.10: Cut excessive wire leads with a wire cutter

Make sure to verify that the connections on your circuit match your schematic, and if needed, use a multimeter's conductivity function to test the board.

Solder a Simple Circuit with Perfboard

Let's try to solder a simple circuit on a perfboard, the components are not mandatory, use whatever you feel comfortable. We have provided a sample diagram for you below:

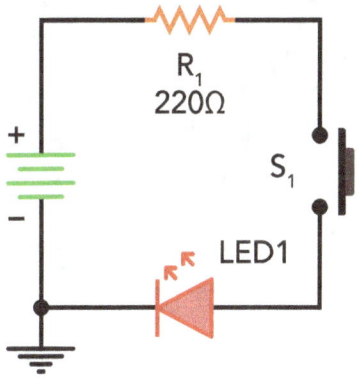

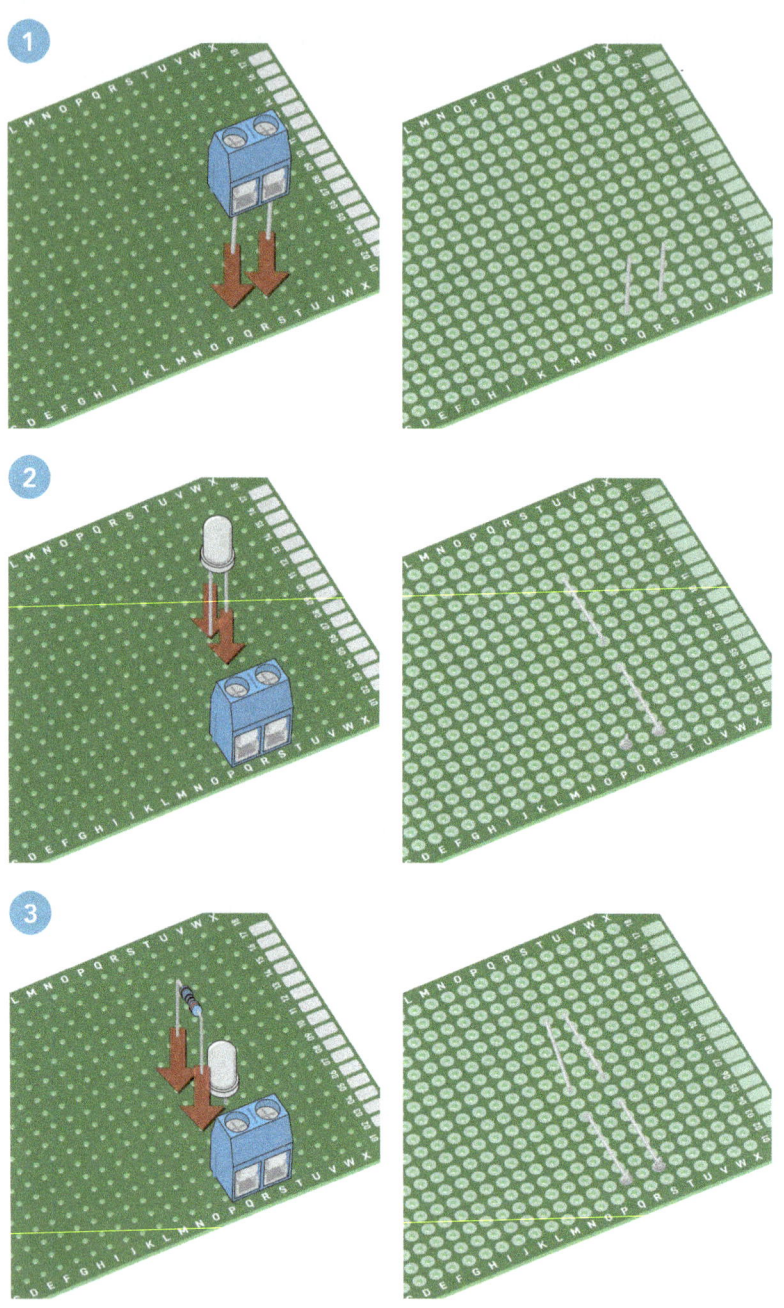

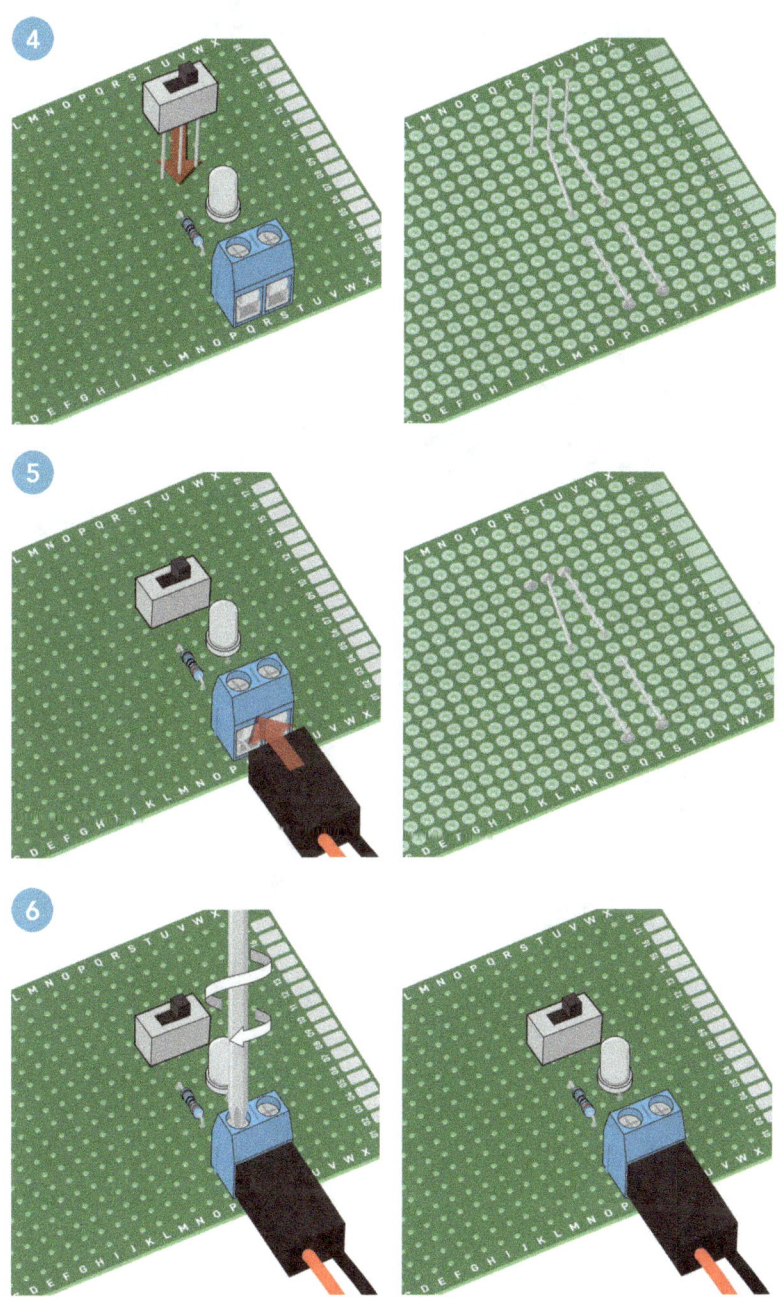

Desoldering THT Components

To remove or replace a component from the board, you must first desolder this component. Typically there are three tools that can help you desolder a THT component: a hot air gun, a desoldering pump or desoldering braid.

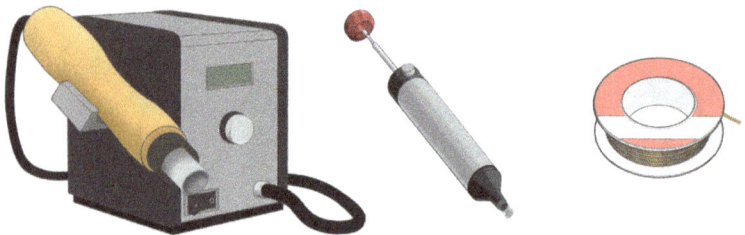

Figure 4.11: Commonly used tools for desoldering

We typically do not recommend a hot air gun for beginners, as it can easily overheat or damage components if used improperly. Here we will show simpler methods such as using a desoldering pump or wick.

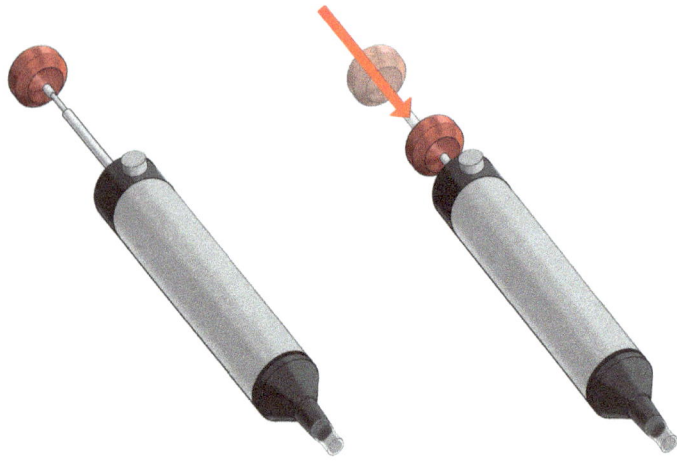

Figure 4.12: A desoldering pump

A desoldering pump is a tool used to remove excess solder from a circuit board. You can use your soldering iron to melt solder from a connection you want to remove — then the desoldering pump can be used to create a vacuum that sucks up that solder.

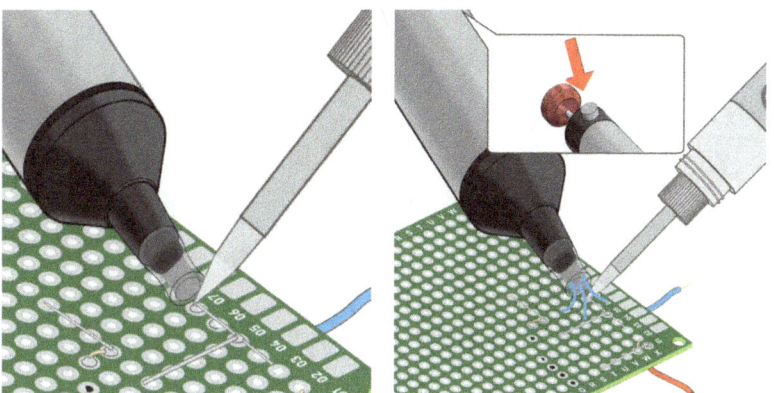

Figure 4.13: Use a desoldering pump to melt and then suck up the liquid solder

Though the nozzle is made of special polymer that can withstand high temperature, you still need to complete the suction process quickly to avoid overheating. The desoldering pump also needs to be cleaned regularly.

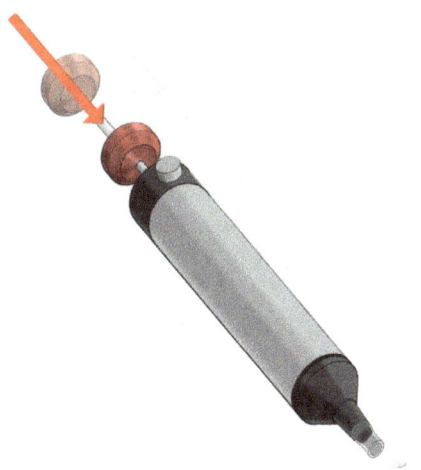

Figure 4.14: Dump clean the waste solder from the pump

Desoldering braid is a fine copper wire with special contexture that can remove solder from electronic components by absorbing and wicking away molten solder.

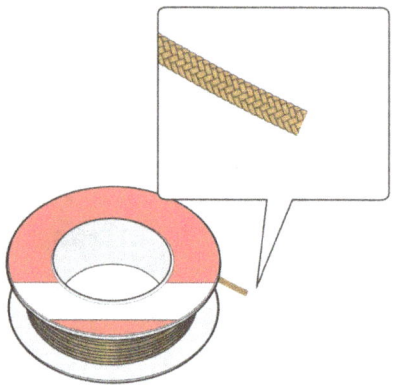

Figure 4.15: The contexture creates stronger grip to the liquid solder

Desoldering wicks may come in different sizes depending on your application. We recommend choosing the wick which which is roughly the same size as your soldering tip or pads on the PCB.

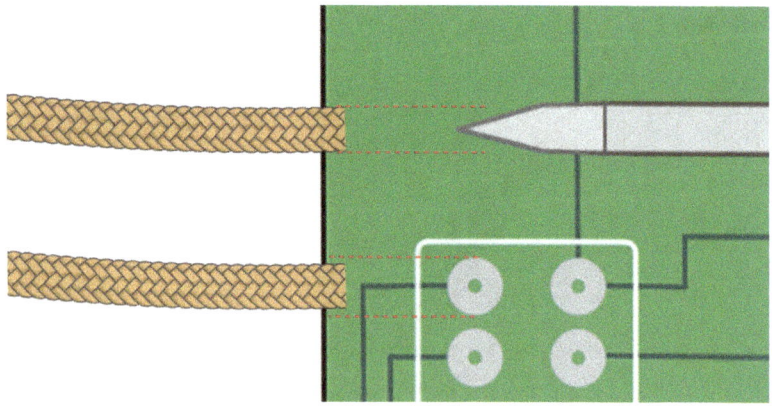

Figure 4.16: Use appropriately sized desolder wick

To begin, we recommend applying some flux onto the wick as the first step. Then use your soldering iron to heat the target you want to remove solder from.

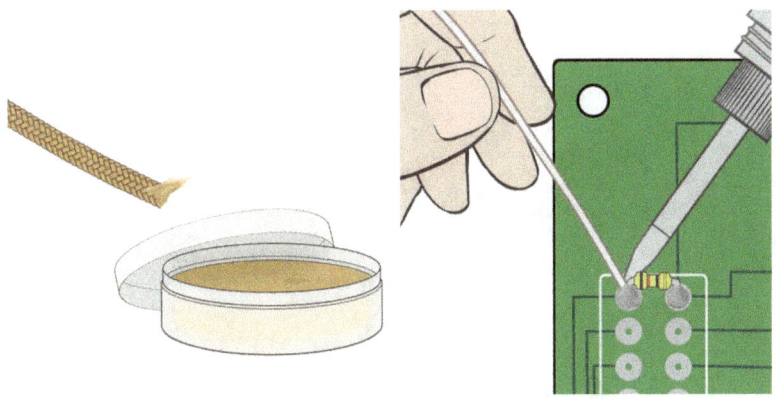

Figure 4.17: Adding additional solder to help accelerate the melting process

As the solder on the pad gets melted, place the desolder wick onto the target and slowly slide it through to pad to allow solder stick onto the wick.

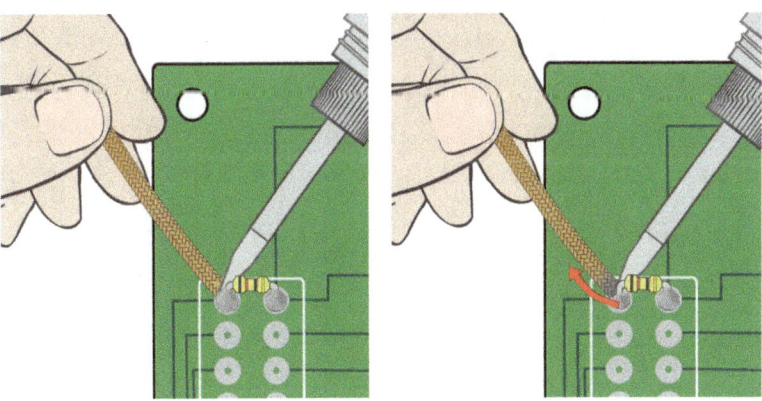

Figure 4.18: Use wick to remove excess solder

Note that the wick is also thermally conductive, so it will get hot and you should keep your fingers some distance away from the tip.

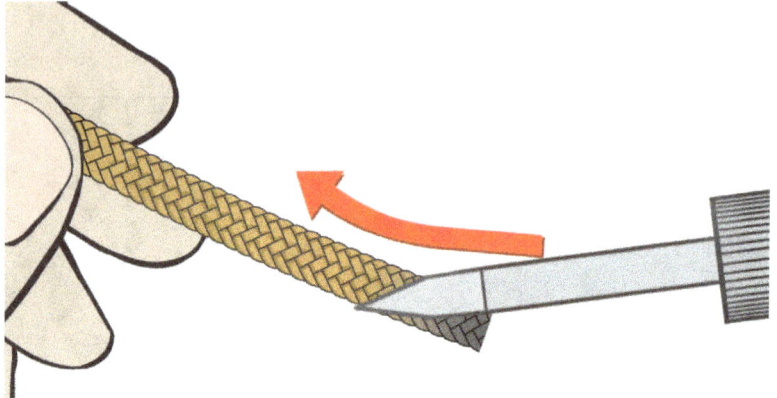

Figure 4.19: The segment of wick with solder attached is not longer reusable

Project - LED Chaser

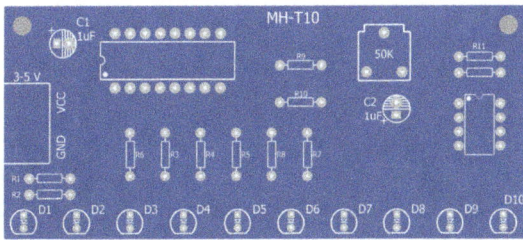

Component List

Name	Spec	Quantity	Position
Resistor	1kΩ	10	R1-R10
Resistor	2.2kΩ	1	R11
Resistor	10kΩ	1	R12
LED	Red	10	D1-D10
Electrolytic Capacitor	1uF	2	C1, C2
Potentiometer	50kΩ	1	R13
Screw Terminal	2 POS	1	J1
SO-08 IC	555	1	U1
SO-16 IC	4017	1	U2

Schematic Diagram

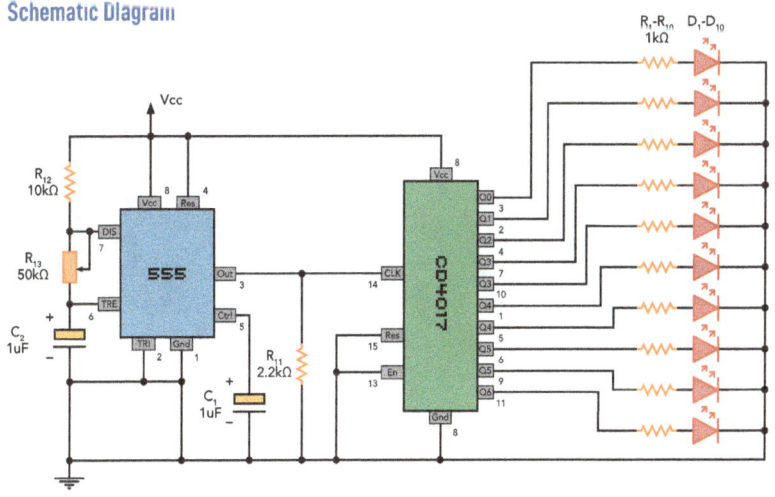

Soldering Order

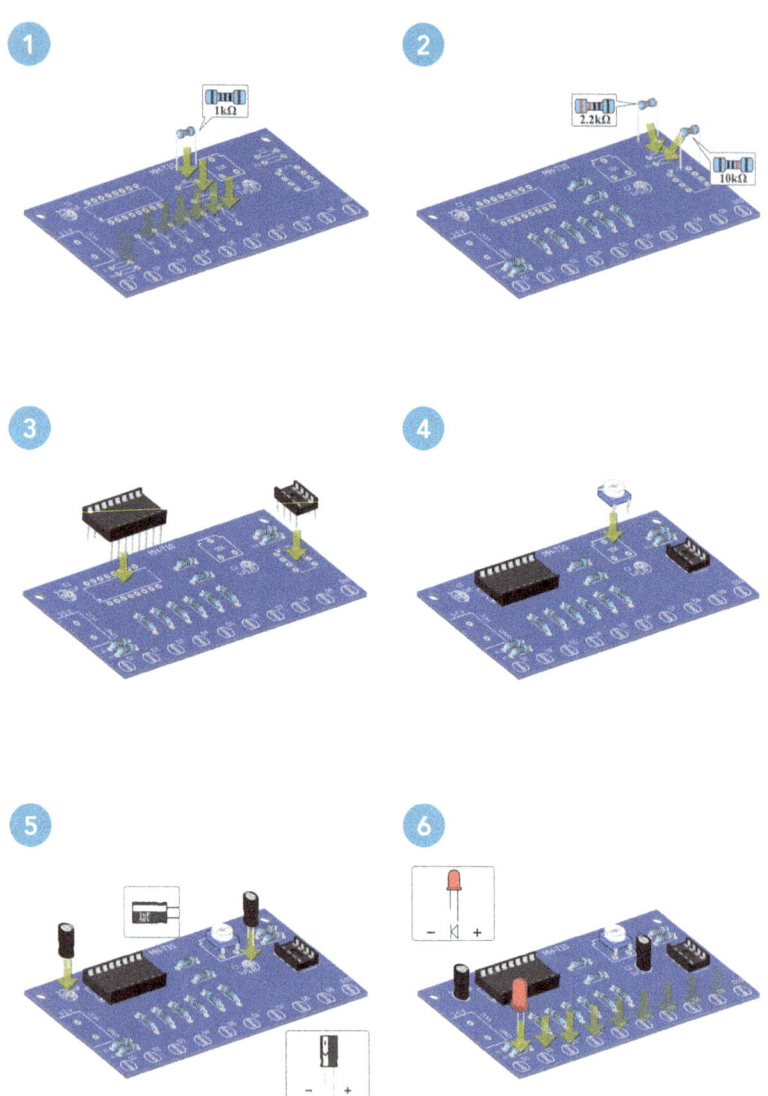

7

8

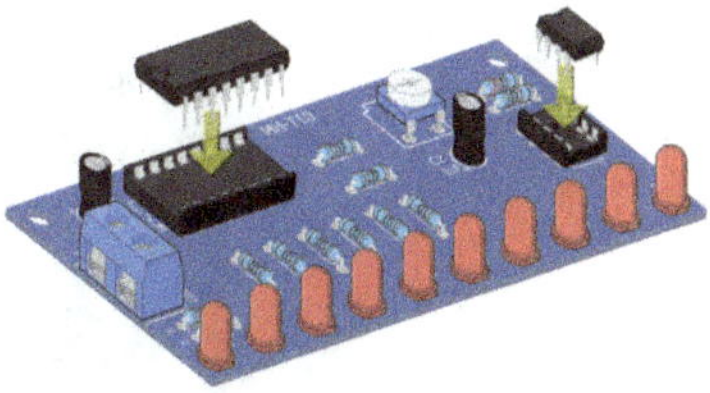

9

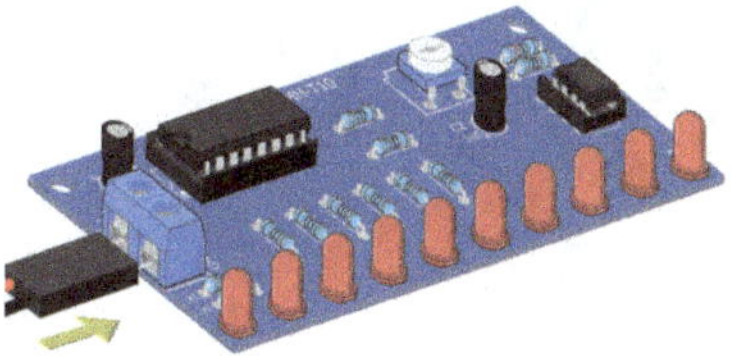

10

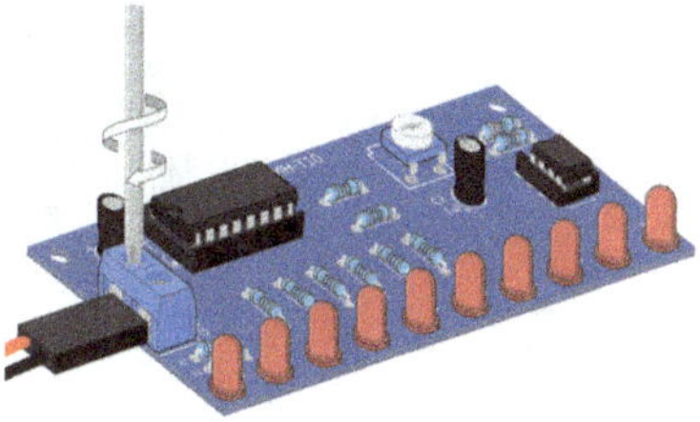

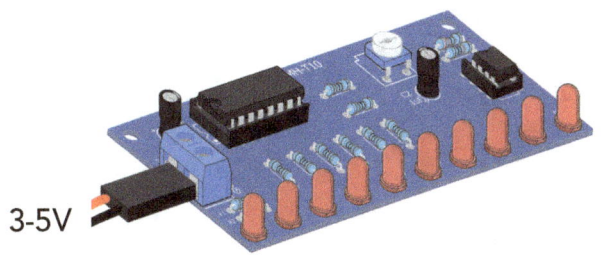

3-5V

Project - Metal Detector

Component List

Name	Spec	Quantity	Position
Resistor	470Ω	1	R3
Resistor	2kΩ	1	R2
Resistor	200kΩ	1	R1
LED	White	1	LED1
BJT	9012	2	Q2, Q3
BJT	9018	1	Q1
Ceramic Capacitor	104	2	C1, C4
Ceramic Capacitor	222	2	C2, C3
Electrolytic Capacitor	100uF	1	C5
Potentiometer	100Ω	1	VR1
Screw Terminal	2 POS	1	J1
Buzzer		1	SP1
Self-lock Switch	DPDT	1	SW1

Schematic Diagram

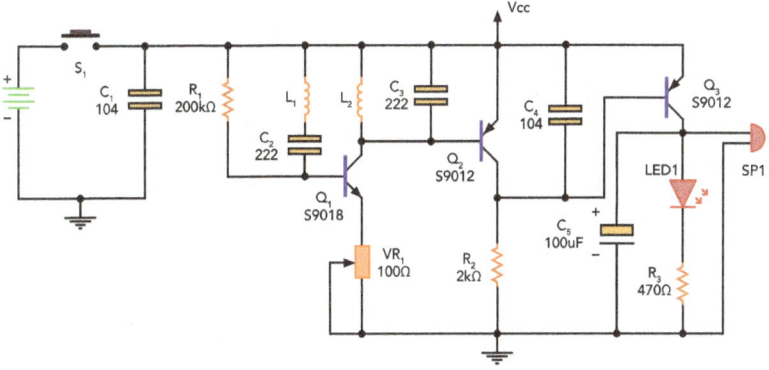

Soldering Order

1

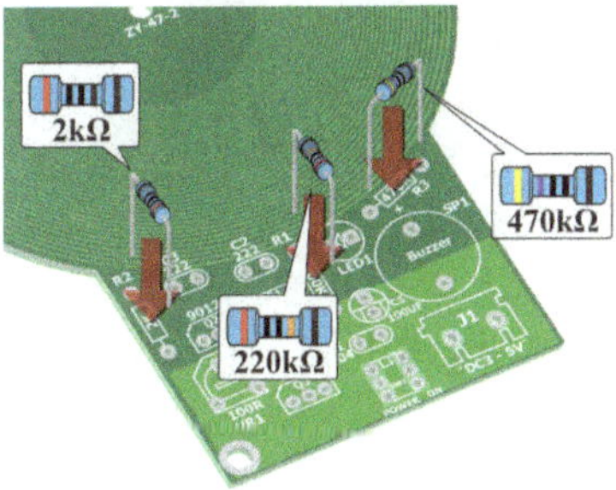

2

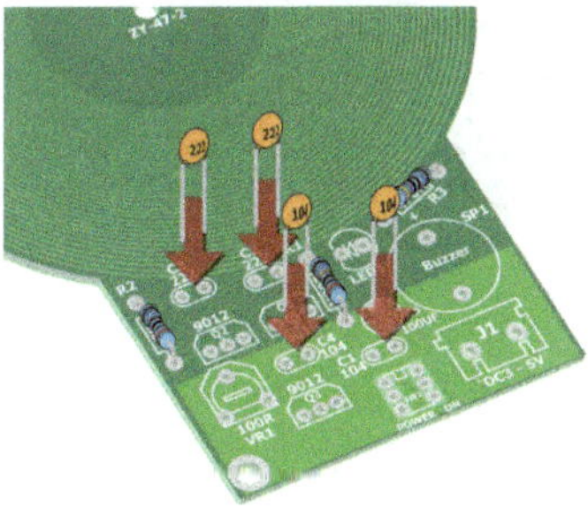

3

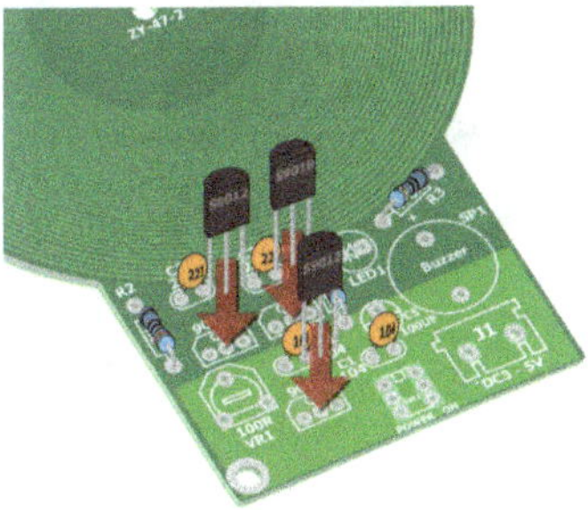

4

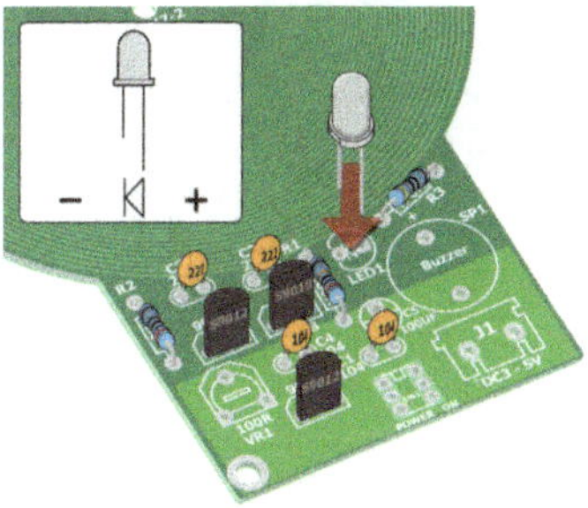

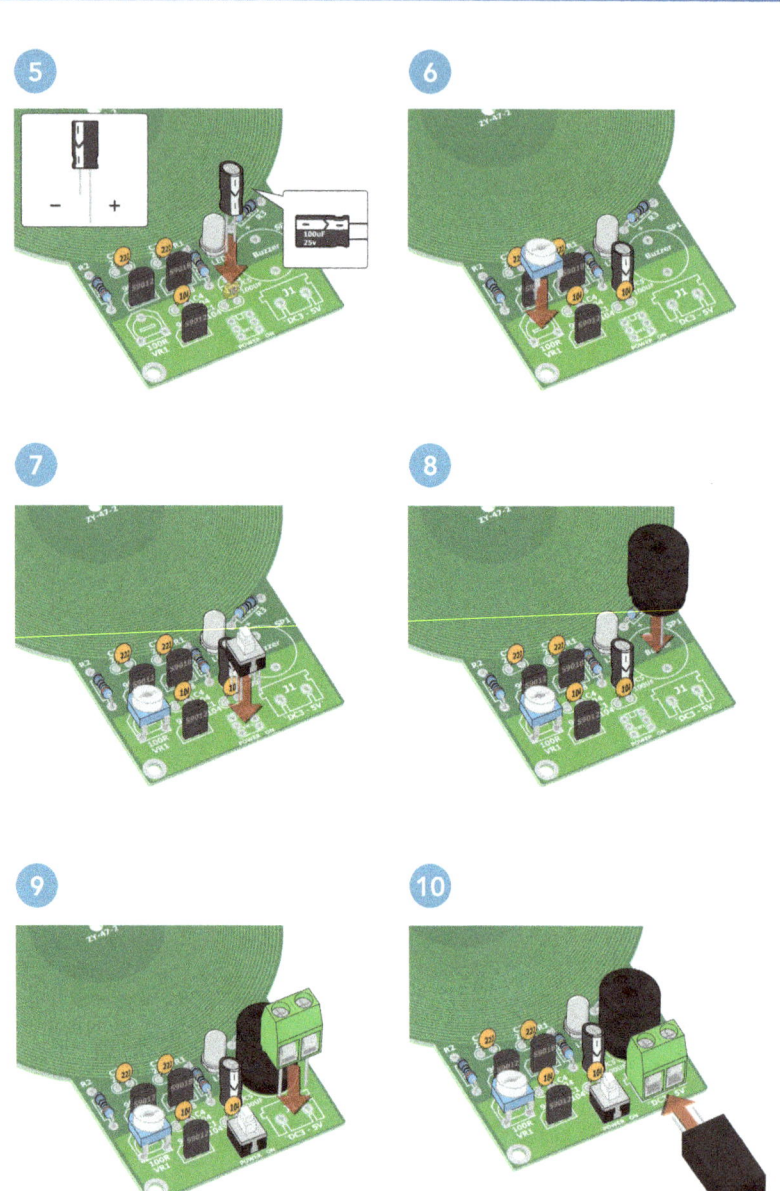

11

12

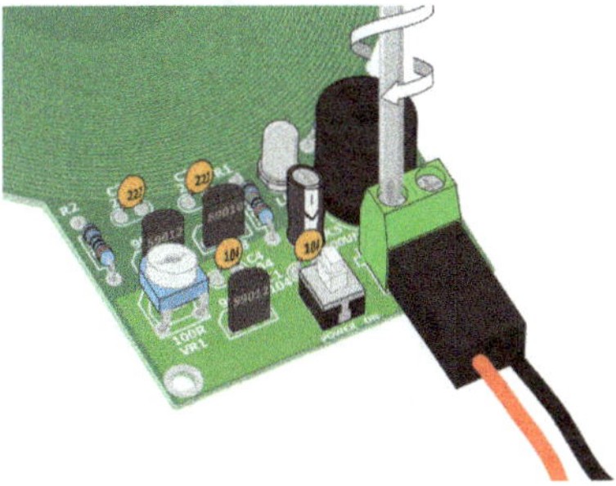

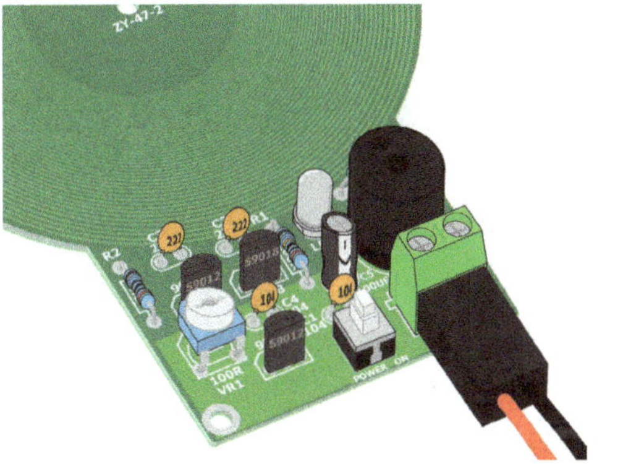

3-5V

Project - Lucky Wheel

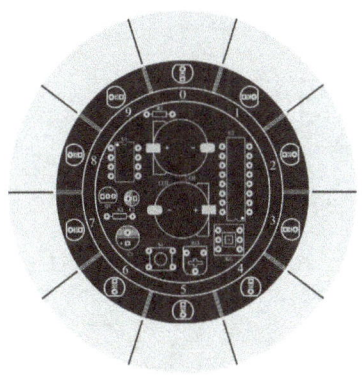

Component List

Name	Spec	Quantity	Position
Battery Holder	CR1220	2	BT1, BT2
Resistor	10kΩ	1	R2
Resistor	470kΩ	1	R3
LED	Red	10	0-9
NPN Transistor	S9014	1	Q1
Electrolytic Capacitor	220uF	1	C1
Electrolytic Capacitor	1uF	1	C2
Potentiometer	1MΩ	1	R13
Switch		1	S1
Self-lock Switch	DPDT	1	KG
SO-08 IC	555	1	U1
SO-16 IC	4017	1	U2

Schematic Diagram

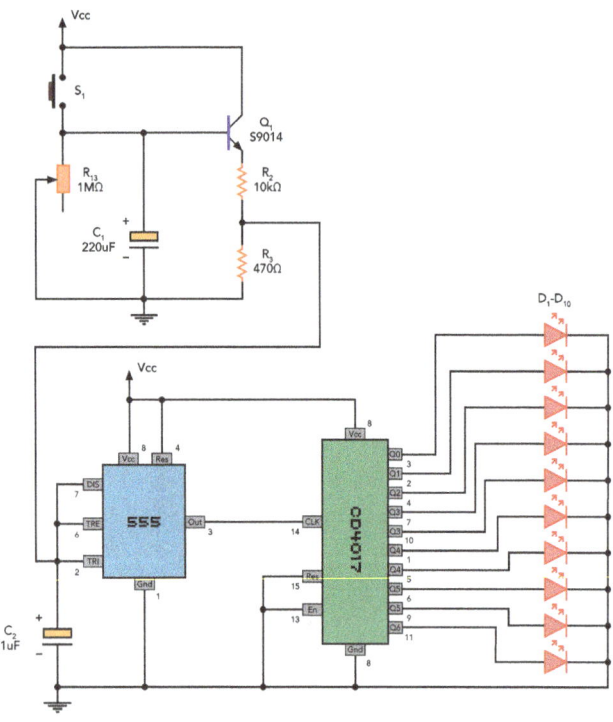

Soldering Order

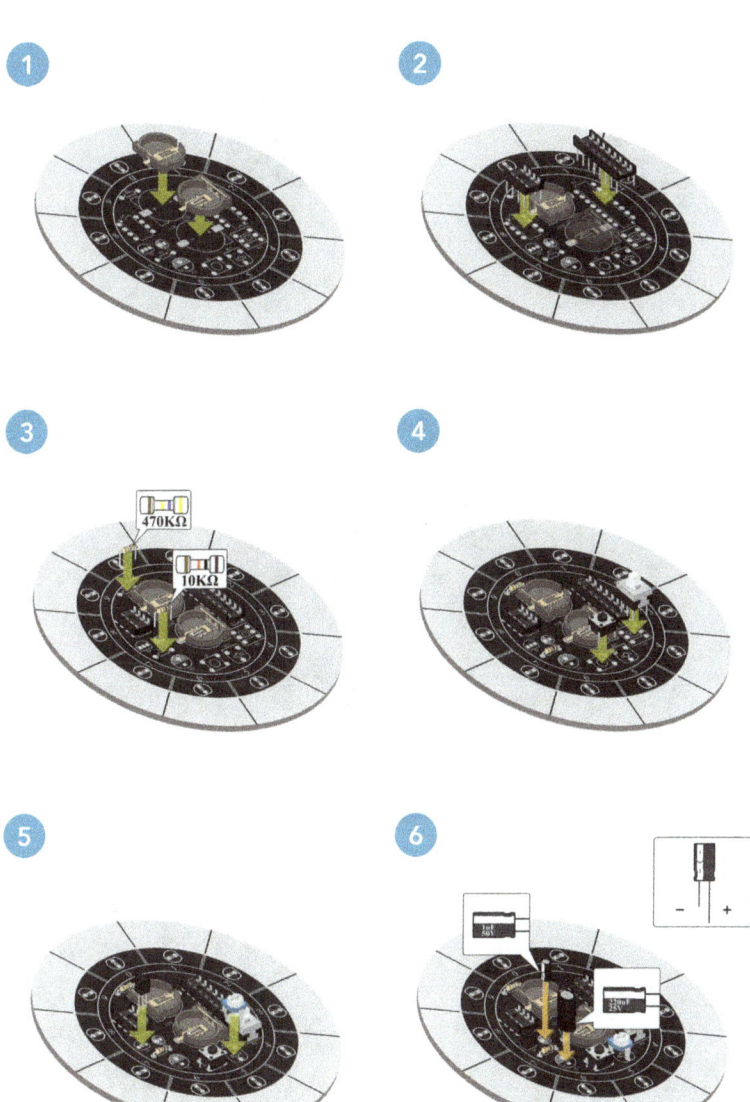

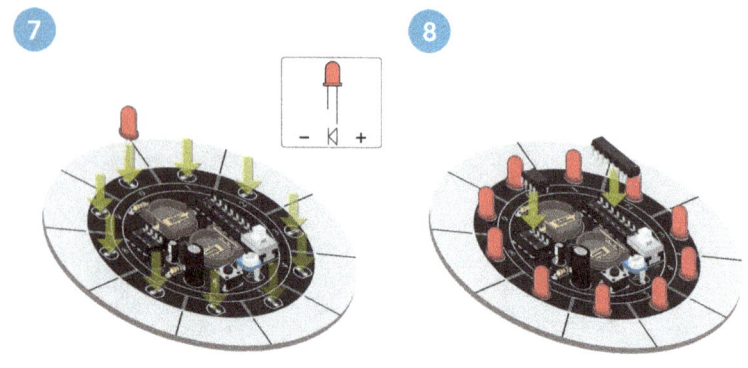

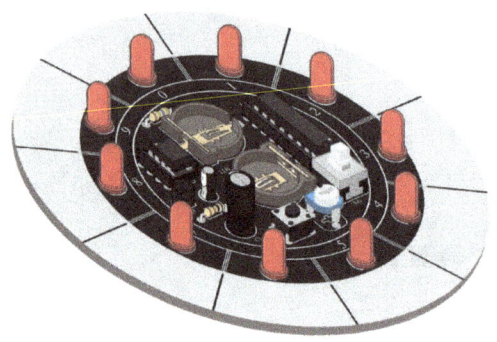

V - Surface Mount Technology Basics

In modern times, electronics products tend to be of a smaller size and have higher component density. As a result, Surface Mount Technology (SMT) components have become the prevalent choice in commercial electronics.

Surface Mount Technology Basics

SMT components can be attached directly onto the physical layout or arrangement of the solder pads on a PCB (known as footprint).

Figure 5.1: Mount an SMT chip to the pads on a PCB

The sizes for SMT components are way smaller than THT parts, allowing higher component densities on boards.

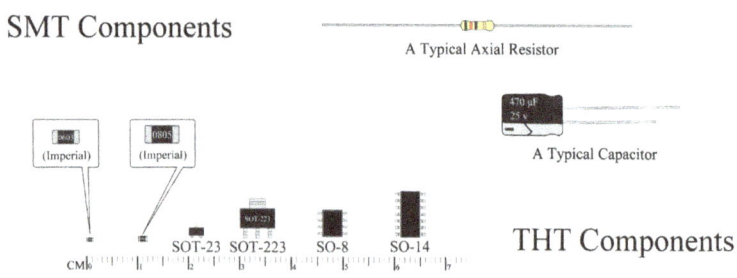

Figure 5.2: Comparison of sizes for SMT and THT components

Prepare for SMT Soldering

Small SMT components like resistors and capacitors are packaged in cut tapes that are covered with a thin film on the top to prevent them from falling out.

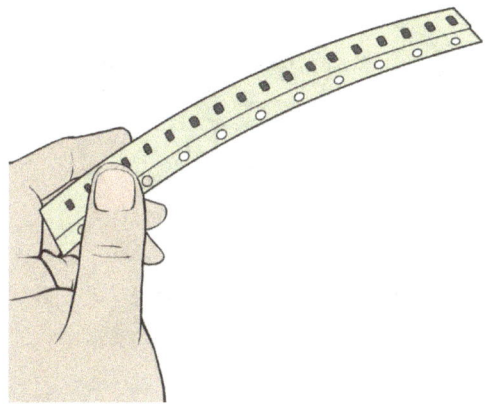

Figure 5.3: SMT components packaged in a segment of cut tape

These parts are small and light which can be easily dislodged even by a slight breeze. Therefore, when handling these parts, stay cautious and gently remove the protective film before placing them in a secure and stable location.

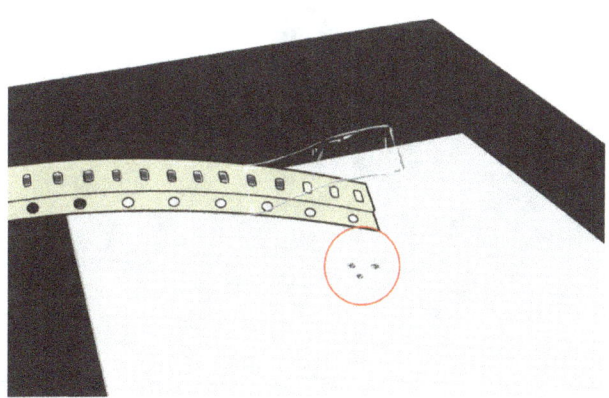

Figure 5.4: Labeling component value on paper also reduces chances of mistakes

Obviously your fingers are too big to pick up and hold these components, so an electronics tweezer is needed to do the job.

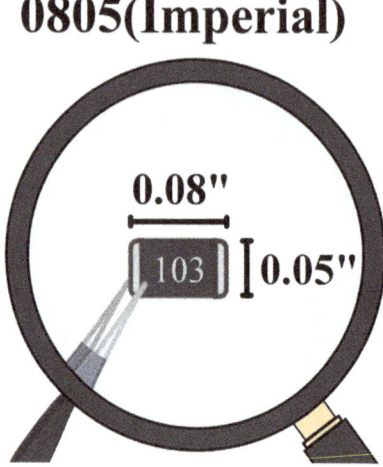

Figure 5.5: A closer look of SMT resistor (0805 footprint, with dimension labeled)

The markings on SMT resistors, though barely visible, represent their values. For example, a resistor marked with 103 means it has a resistance of 10k Ohms.

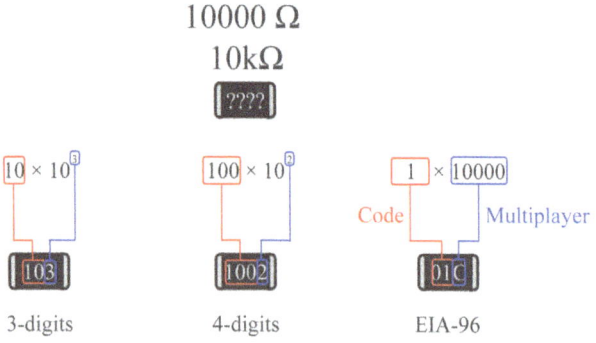

Figure 5.6: Interpreting SMT resistor markings

In the first section we mentioned different types of soldering tips. While soldering small SMT components, you will find conical tips are easier to use. Meanwhile, you want to use solder wires with thinner dimensions.

Soldering SMT with Two Terminals

SMT components are mainly seen on PCBs, which may have many components and pads next to each other. Make sure you choose the right component and check against PCB layout diagram to ensure accurate placement.

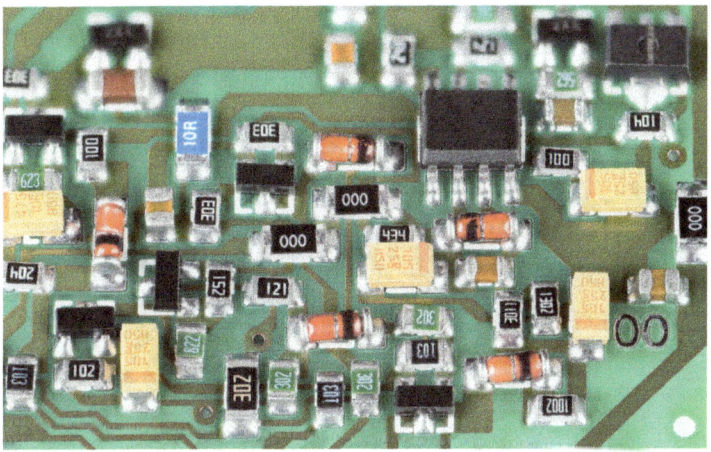

Figure 5.7: Terminal pins of SMT components can vary from two to hundreds

Resistors, capacitors and diodes are two terminal components, so the soldering process is relatively easy. For example, to solder a resistor onto the board, we start off with soldering onto one of the pads on board.

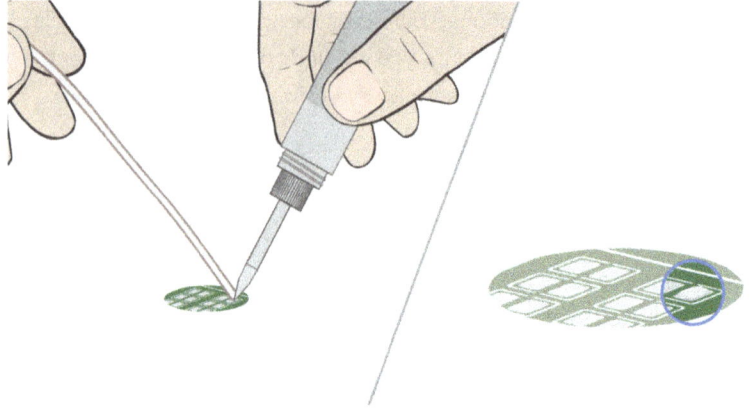

Figure 5.8: The first step is to apply solder on one pad

Then, use a tweezer to hold the component in place. Solder one pad with one pad to one leg of the resistor. Once the solder is solidified, use the tweezers to slightly push the component and see if it has been securely attached to the pad.

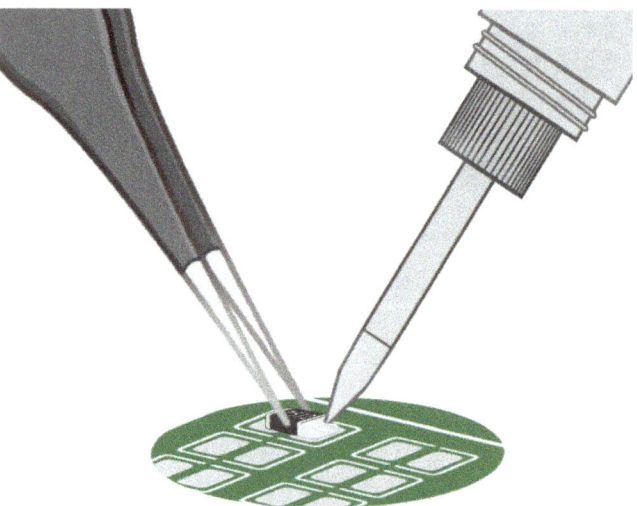

Figure 5.9: Use tweezers and soldering iron to hold one terminal onto the pad

Since one leg of the resistor has been secured in position, you can simply apply solder to the other leg to create the electrical connection to the board.

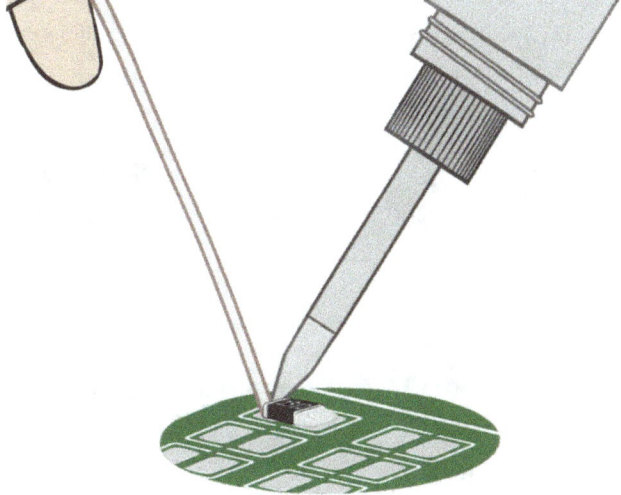

Figure 5.10: Apply solder on the other side to create electrical connection

If you are soldering polarized components such as diodes, electrolytic capacitors or LEDs, make sure their polarities match the marking on the board.

Soldering More Complicated Components

Transistors and integrated circuits (ICs) have more than two pins. They have more complicated SMT footprints such as SOP (Small Outline Package), SOJ (Small Outline J-Lead), TSOP (Thin Small Outline Package), SOIC (Small Outline Integrated Circuit), and others.

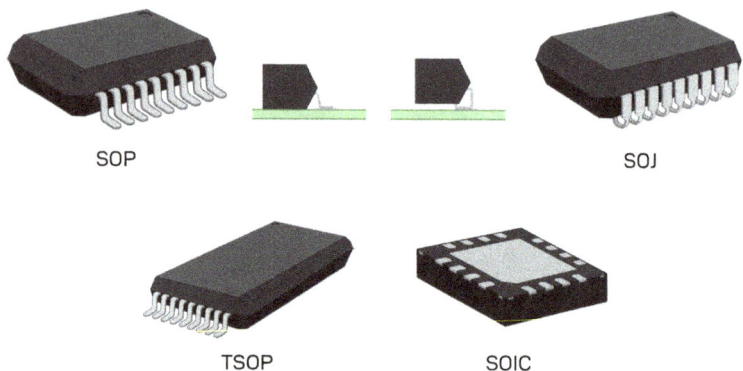

Figure 5.11: SMT integrated circuits (ICs) usually contain more terminals

Just so you know, the common industrial approach for mounting SMT components with multiple pins involves applying solder paste onto the pads, which is typically accomplished using PCB stencils, or can be done manually in simpler cases.

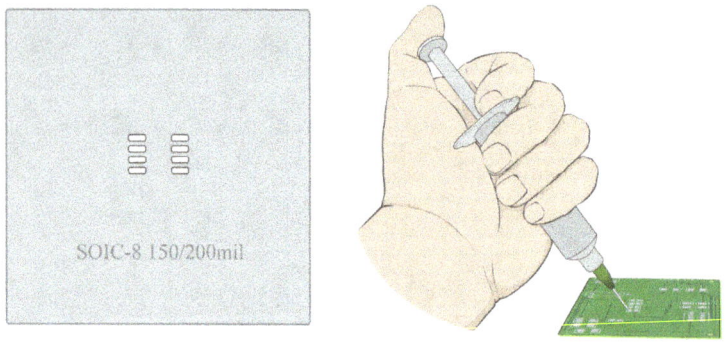

Figure 5.12: Applying solder onto pads using a stencil or manually

Then the IC needs to be aligned and placed on top of the pads with solder paste. This process can be quickly completed with an industrial chip mounter machine. For manual soldering, we will use a tweezer to place the component onto the pads.

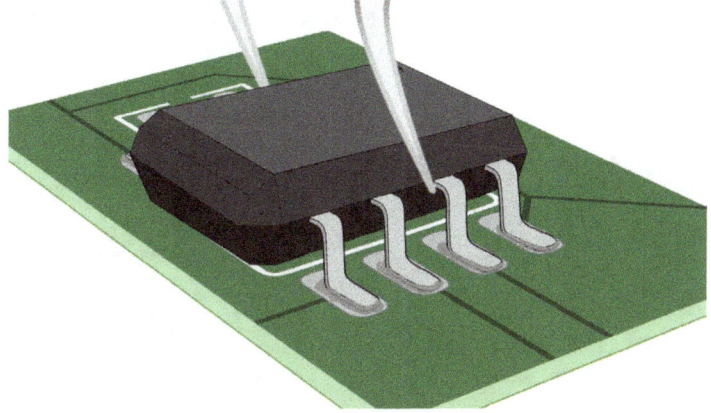

Figure 5.13: Manually place the chip onto the board using tweezers

To create a permanent bond between the parts and the board, we need to follow a proper heating and cooling process, which is usually done with a reflow oven that precisely controls the environmental temperature.

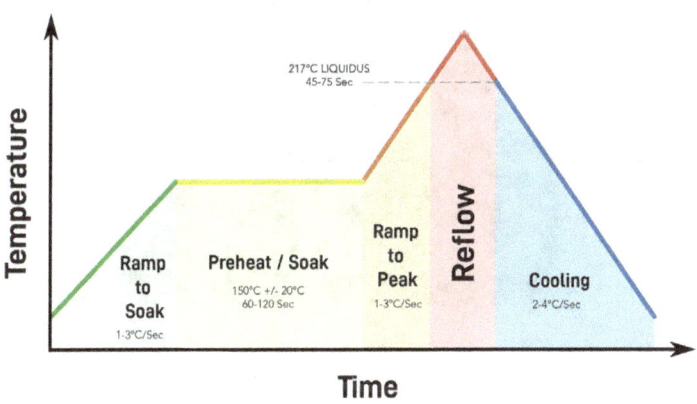

Figure 5.14: Typical temperature control curve in an industrial reflow oven

You can also try to heat the board manually with a hot air gun, but the problem is that the IC can be overheated if not done properly.

Industrial production aside,, for our DIY projects or quick testing & debugging, we still recommend using a soldering iron to solder SMT ICs. For example, to solder an SOP-8 component, you always start with tinning one of the pads first.

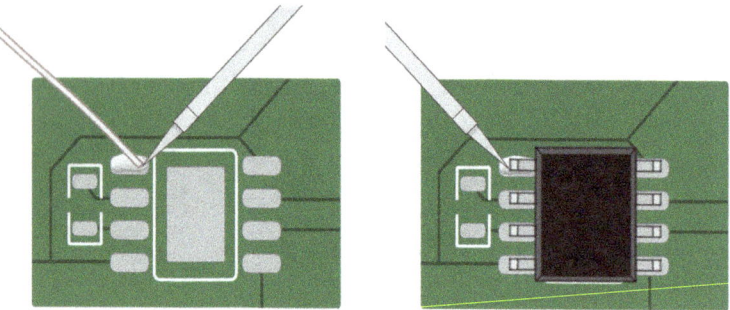

Figure 5.15: Manual soldering of SMT ICs also starts off with one pad in the corner

Then we need to position the IC with a tweezer while utilizing this pad to lock the position.

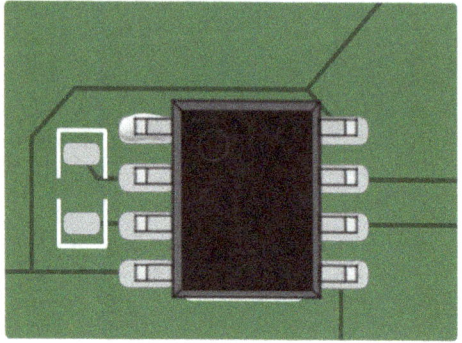

Figure 5.16: Once soldified, it creates a firm bond that fixes the position of the chip

Now we will need to solder the rest of the pins properly just like we learned previously.

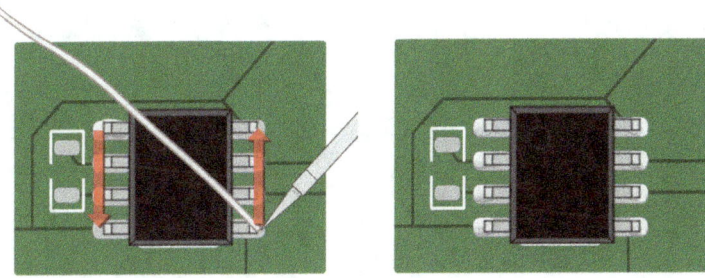

Figure 5.17: Solder all the rest pins and make sure they are electrically connected

ICs with denser pins (e.g. SOIC, QFP) are trickier to work up. As pins getting close to each other, you may create some unwanted solder bridges, which can cause malfunction of the circuit and should be totally avoided.

Figure 5.18: Always make sure the solder bridges are removed

Again you will need flux to help with that. Using an SOIC28 footprint as an example, we will first apply some solder flux on the two corner pads.

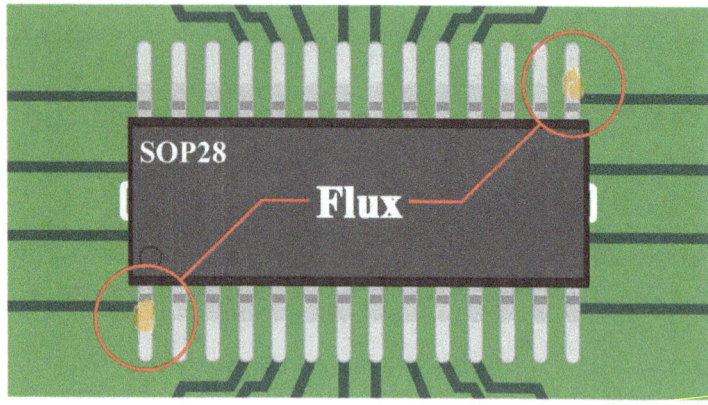

Figure 5.19: Solder the corner pins first (using flux makes the process easier)

Apply some solder onto the soldering iron and solder the two terminal pins. Do not worry even if you have applied excessive solder on these two pins since the only purpose for this step is to mount and hold the position of the chip.

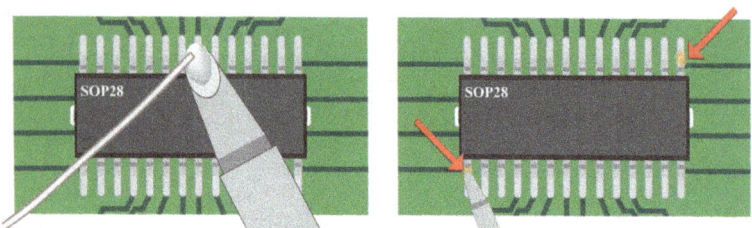

Figure 5.20: Flux makes solder more easily bonded onto the pins

To work on the pins in the middle, apply flux evenly onto the pin first.

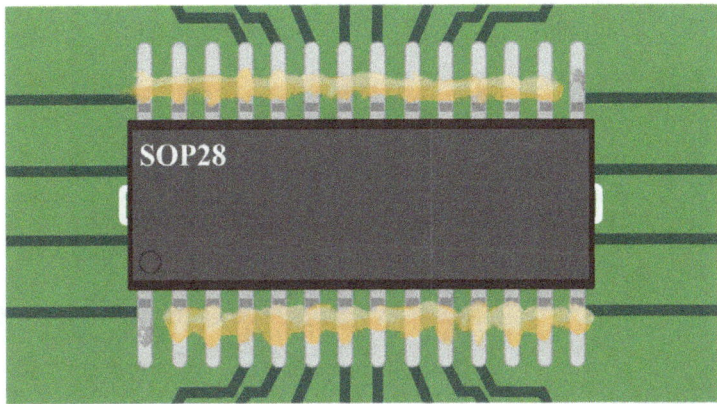

Figure 5.21: Use a toothpick to apply the paste-like rosin flux onto the chip pins

The next step is call drag soldering, which is an advanced and commonly used skill in SMT soldering.

Figure 5.22: The flux on the pins enables solder to be dispensed onto each pin

If solder bridges occurred, clean the tip and target bridged pins by (explain what you mean by short-dragging) a few times to eliminate the solder bridge. This is a very tricky skill that will take practice to master.

Figure 5.23: A perfectly soldered SOIC 28 chip

Practice Board

Before diving into real SMD projects, we strongly advise beginning with our SMD practice board. It's the perfect platform to hone your soldering skills. Experiment with various component sizes until you're at ease with SMD soldering. Once you're totally comfortable with a practice board, you can confidently progress to your next project.

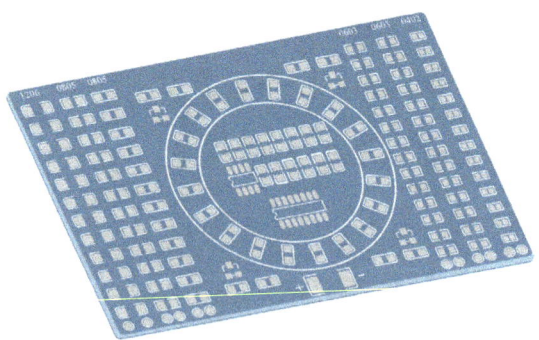

Figure 5.24: Practice board included in our kit

Component List

Name	Spec	Quantity	Position
0805 Resistor	102	17	R50-R60, R65-R68
0805 Resistor	103	6	R48, R61-R64
0805 Resistor	106	2	R49
0805 Capacitor	0.01uF	3	C27, C28
0805 LED	Red	12	D1-D11
0805 LED	Blue	4	D16-D19
LL34 Diode	4148	4	D12-D15
SOT23 Triode	J3Y	4	Q1-Q4
SO-08 IC	555	1	U1
SO-16 IC	4017	1	U2
1206 Resistor	N/A	14	R1-R12
0805 Capacitor	N/A	14	C1-C12
0805 Resistor	N/A	14	R13-R24
0603 Capacitor	N/A	16	C13-C26
0603 Resistor	N/A	16	R34-R47
0402 Resistor	N/A	16	0402 Column

Structure Diagram

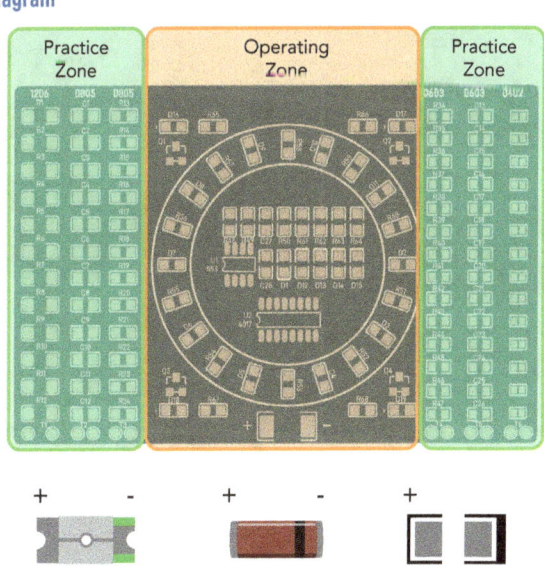

LED   4148 Diode   PCB Mark

Schematic Diagram – Operating Zone

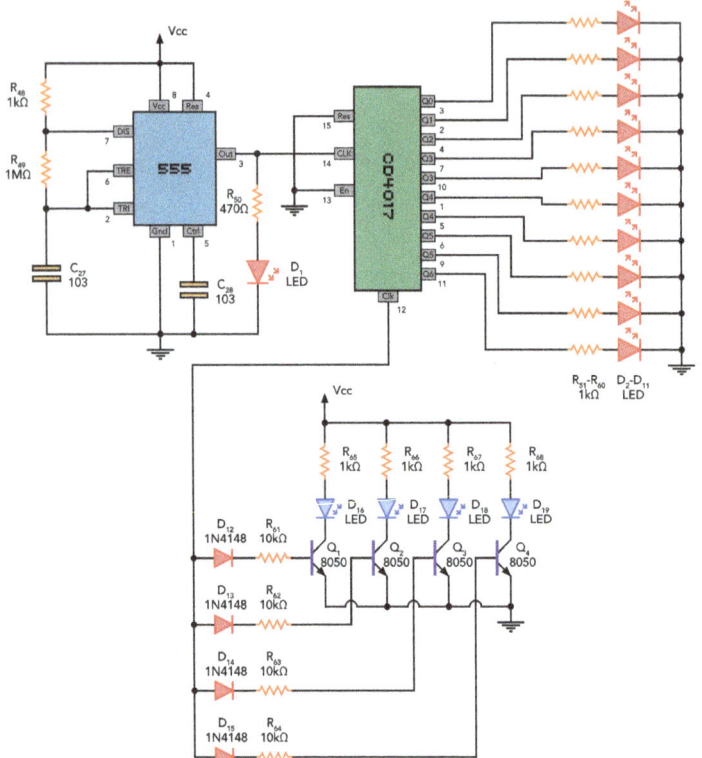

Project - Fidget Spinner

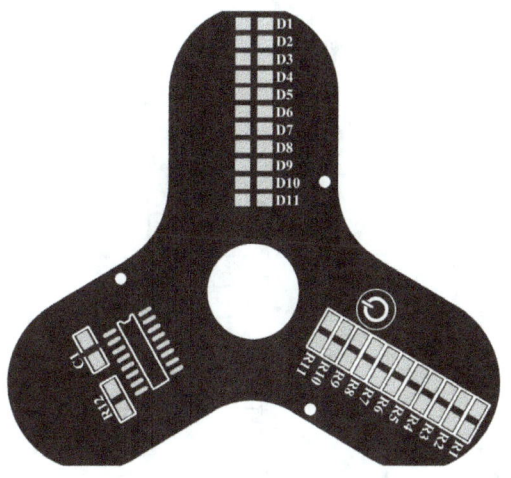

Component List

Name	Spec	Quantity	Position
Battery Connector	CR1620	3	BT1, BT2, BT3
0805 Resistor	101	12	R1-R11
0805 Resistor	103	2	R12
0805 Capacitor	104	1	C1
0805 LED	Red	6	D1-D11
0805 LED	Green	6	D1-D11
0805 LED	Blue	6	D1-D11
Switch		1	MS
SO-14 Microcontroller		1	U1
Self-tapping screws		4	
Bearing		1	

Schematic Diagram

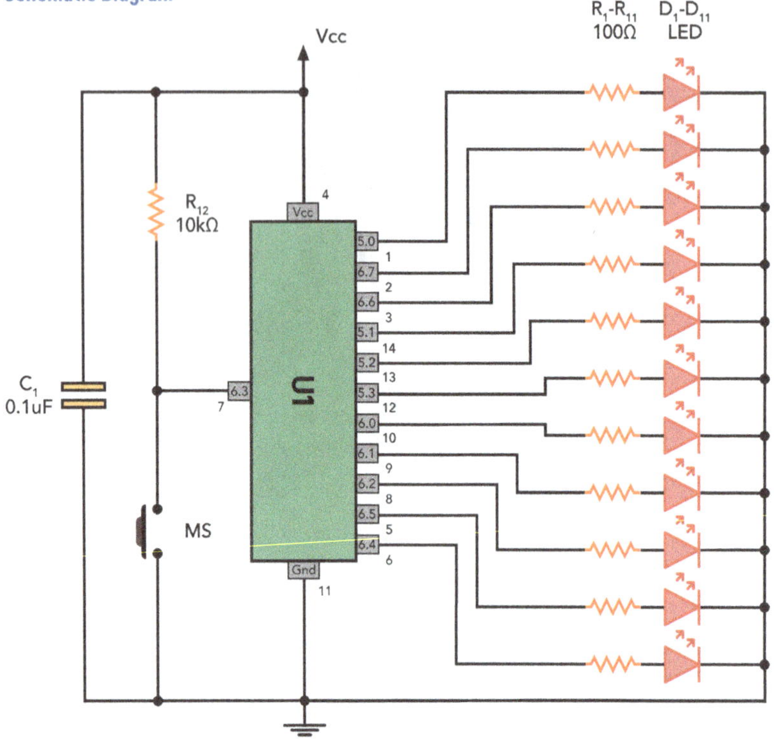

Vcc

R₁-R₁₁ 100Ω D₁-D₁₁ LED

R_{12} $10k\Omega$

4

Vcc

6.0
6.7
6.6
5.1
5.2
5.3
6.0
6.1
6.2
6.5
6.4

1
2
3
14
13
12
10
9
8
5
6

7 6.3

U1

C_1 $0.1uF$

MS

Gnd

11

Installation Order

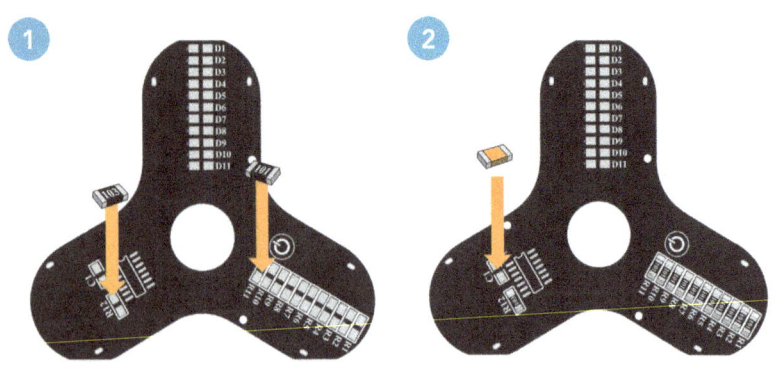

1

2

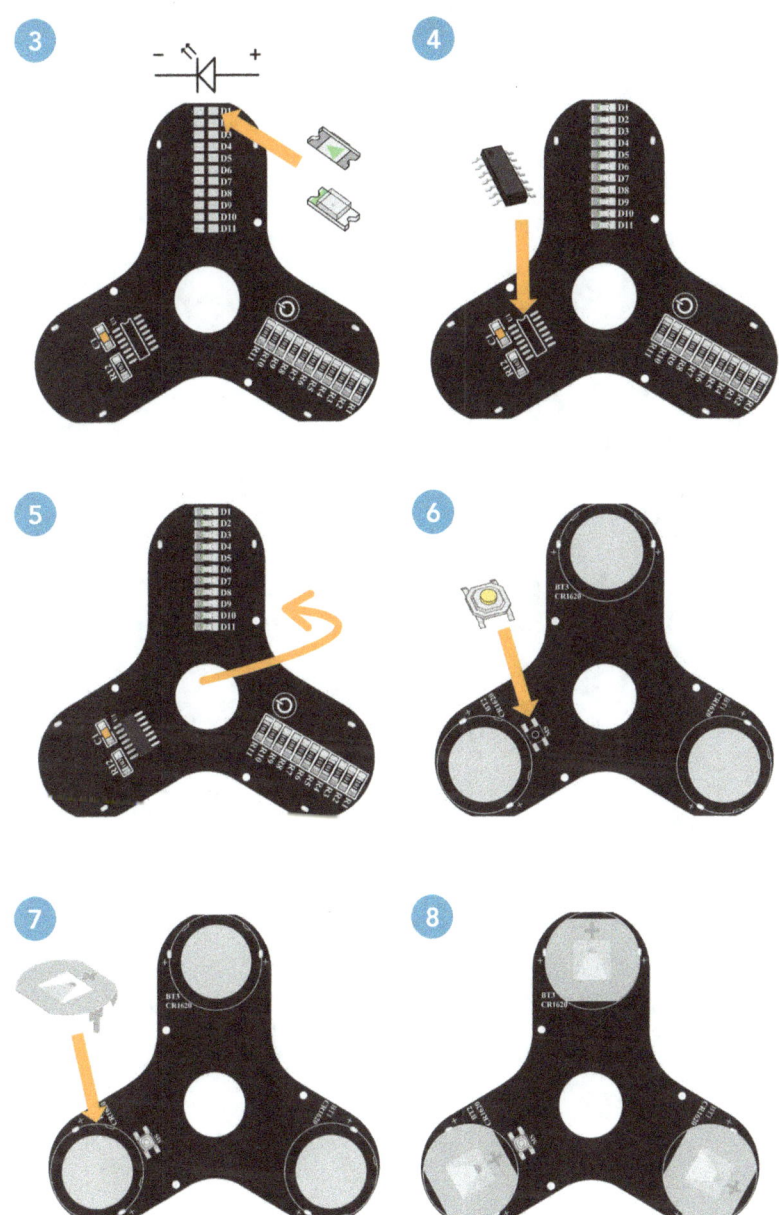

APX 1 – Common Mistakes to Avoid

Soldering may seem intimidating, but it's far from rocket science. In fact, it's a fun and beginner-friendly activity. With practice, it becomes an easy skill to master. Don't be discouraged by common mistakes—embrace the learning process. Grab a soldering iron, dive in, and discover the joy of creating and connecting with precision.

The general characteristics of a good solder joint:

- Good and completed wetting

- A concave fillet

- Shiny and clean

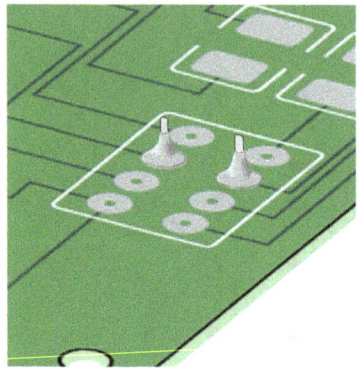

Figure a1.1: An ideal through-hole solder joint

Solder Bridging

Solder bridging occurs when too much solder is applied or the soldering iron tip is too large, connecting unintended solder points and potentially causing short circuits. It can be resolved by melting the solder bridge with a soldering iron or using a desoldering pump to remove excess solder.

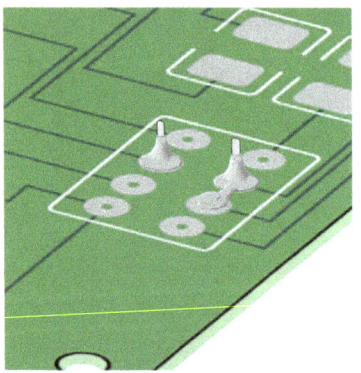

Figure a1.2: Solder bridging

Excessive Solder

Excessive solder application won't impact performance, but it can lead to solder bridging. It can be mitigated with practice finding the right amounts of solder to use. A concave surface is ideal for better observation of the solder joint, and knowing when to withdraw your solder is crucial to avoid excess.

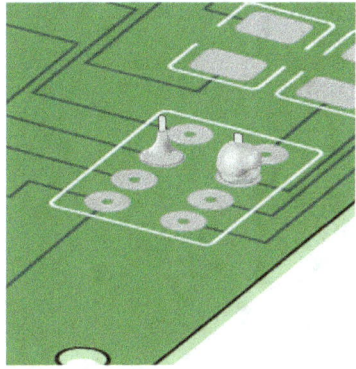

Figure a1.3: Excessive solder

Balling

Solder balling, common defects in wave or reflow soldering, appear as small spheres adhering to laminate, resistor, or conductor surfaces. Moisture near through holes can cause vapor to extrude solder, creating balls on the front side of printed boards. Improper process parameters in wave soldering can also lead to irregular solder balls on the board surface.

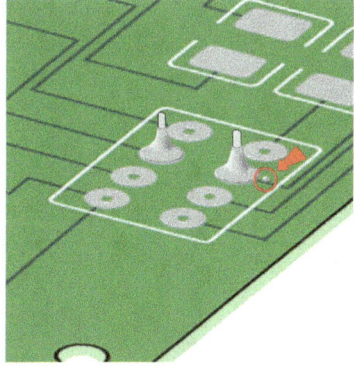

Figure a1.4: Balling

Cold Joint

A cold joint will form an uneven surface, and sometimes partially melted solder can be seen. This issue usually occurs when the solder or soldering iron tip is not adequately heated. Lead-free solder may also be a cause of this problem since it generally has a higher melting point.

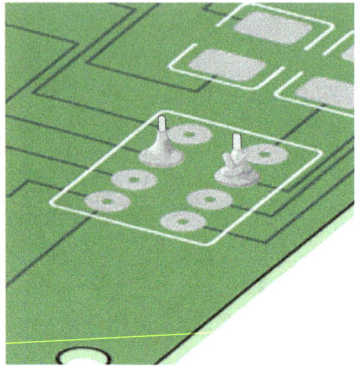

Figure a1.5: Cold joint

Overheated Joint

If the soldering iron tip stays in place for too long, it can lead to overheated joints. This can cause the pad to detach from the PCB, resulting in damage to the entire circuit. Choosing the appropriate temperature can effectively prevent this problem.

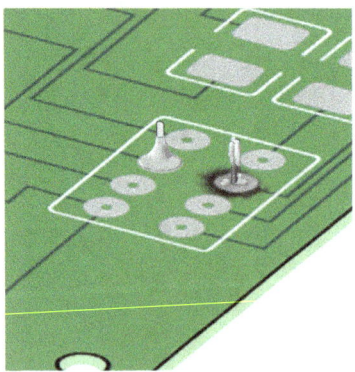

Figure a1.6: Overheated joint

Insufficient Wetting

Insufficient use of flux can prevent the solder from adequately encapsulating the component pins, leading to poor contact. It is important to practice and master the appropriate amount of flux to ensure proper soldering.

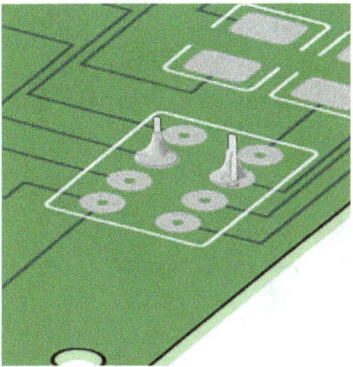

Figure a1.7: Insufficient wetting

Lifted Pads

Lifted pads may occur due to excessive force or overheating during the soldering process. Pad detachment results in a fragile PCB and potential damage.

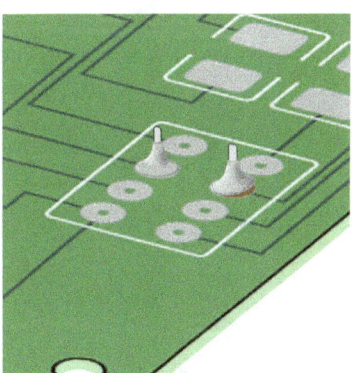

Figure a1.8: Lifted pads

Solder Starved

A solder-starved joint lacks sufficient solder to establish a reliable electrical connection. This issue can be caused by poor solder fluidity, premature solder withdrawal, insufficient flux, or inadequate soldering time.

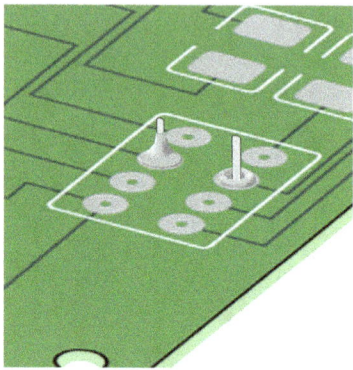

Figure a1.9: A solder starved joint

In SMT soldering, a good joint exhibits the following characteristics:

- Complete wetting, ensuring proper adhesion

- Concave fillet formation, indicating sufficient solder flow

- Shiny and clean appearance, indicating proper soldering and absence of contaminants

Figure a1.10: An ideal SMD solder joint

Tombstoning

If the flux distribution is uneven on both ends of the component, it can cause one end to lift up. In severe cases, one end may not make contact with the pad at all. It is necessary to apply flux evenly to both ends to avoid this phenomenon.

Figure a1.11: Tombstoning

Solder Skips

Sometimes, it just happens.

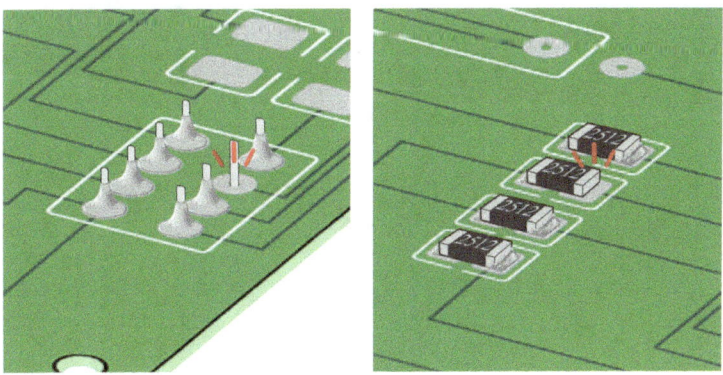

Figure a1.12: Solder Skips

APX 2 - Components and Footprints

Explore the fundamental components found on PCB boards and familiarize yourself with their corresponding footprints on PCB. This section will equip you with the knowledge to identify and work with these essential elements, enhancing your skills in electronics assembly and soldering.

SMD Dimensions

In this section, we will explore key measurements such as length (L), width (B), height (H), and lead spacing (A). These dimensions play a vital role in selecting and correctly placing SMD components on PCBs.

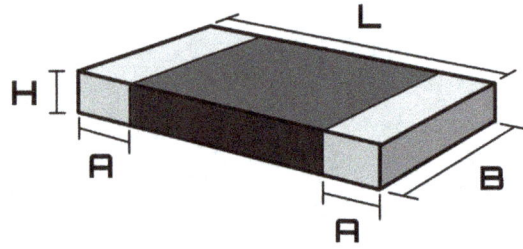

Imperial				Metric		
Case Code	L (in)	B (in)	Power (W)	Case Code	L (mm)	B (mm)
01005	0.016	0.008	0.031	0402	0.4	0.2
0201	0.02	0.01	0.05	0603	0.6	0.3
0402	0.04	0.02	0.062	1005	1.0	0.5
0603	0.06	0.03	0.10	1608	1.6	0.8
0805	0.08	0.05	0.125	2012	2.0	1.25
1206	0.125	0.06	0.25	3216	3.2	1.6
1210	0.125	0.10	0.5	3225	3.2	2.5
1812	0.18	0.125	0.75	4532	4.5	3.2
2010	0.20	0.10	0.75	5025	5.0	2.5
2512	0.25	0.125	1	6332	6.3	3.2

Figure a2.1: SMD Dimensions

Common Components & Footprints

It's essential to familiarize yourself with common components. While you don't need to grasp the intricacies of each component's functionality, understanding key points is vital. This section highlights important considerations when soldering components such as resistors, capacitors, diodes, and ICs, ensuring successful solder joint formation.

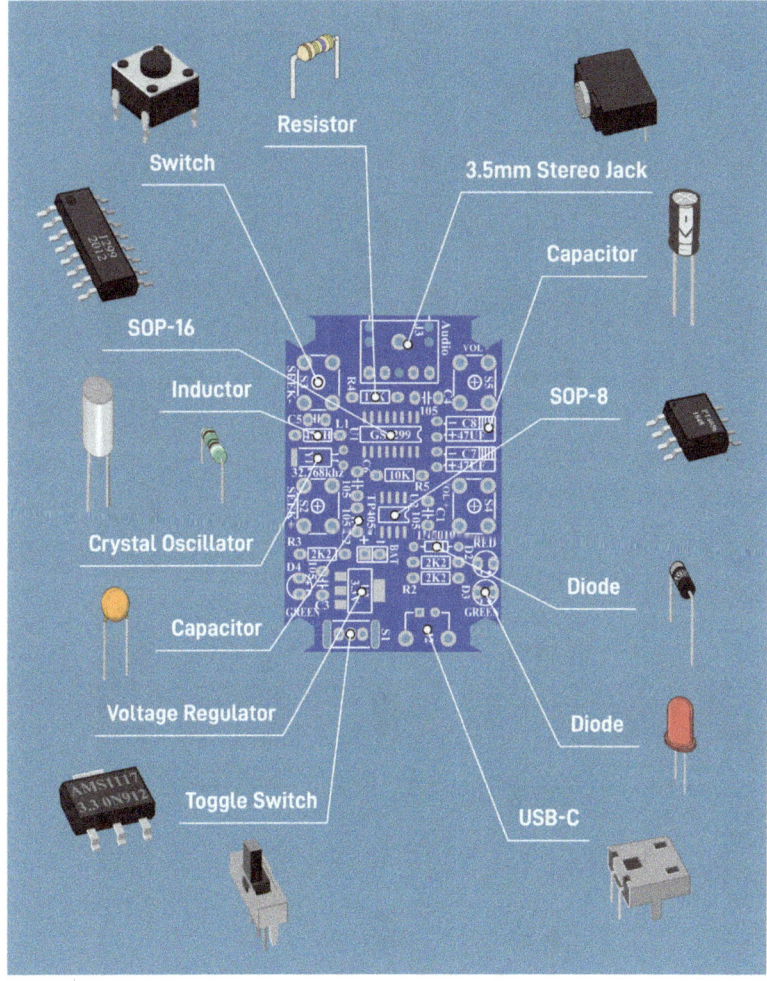

Figure a2.2: Common components on a PCB board

Name	Footprint	Image	Note
Resistor			
Capacitor			The marked side is negative polarity
Diode			The longer pin is anode
			The marked side is cathode
BJT			Line up the front flat side with the footprint on the board

Name	Footprint	Image	Note
Potentiometer			
Voltage Regulator			
SOP-8			Line up the notch or printed dot on the IC with the one on the board.
SOP-16			
3.5mm Jack			
Button			
Toggle Switch			

A multilayer PCB Stackup

A multilayer PCB stackup is comprised of multiple layers of copper traces, insulating materials, and internal planes. It includes signal layers for routing signals, power and ground planes for power distribution, and additional layers for signal integrity. Prepregs and solder mask layers provide insulation and protection. The stackup ensures optimized electrical performance, impedance control, and efficient routing for complex circuit designs.

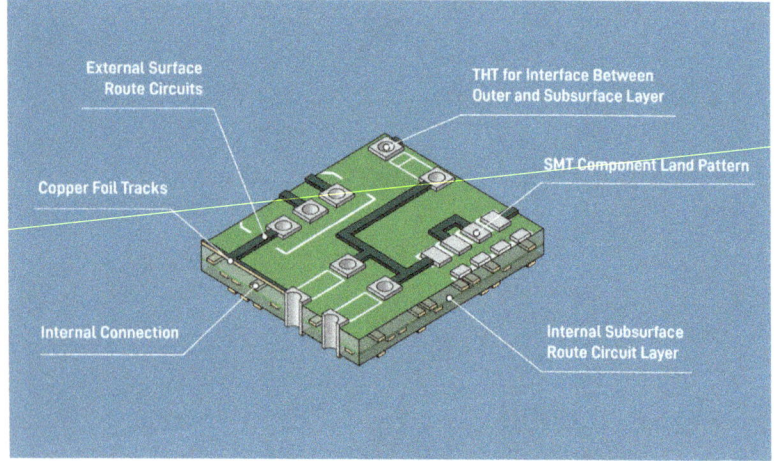

Figure a2.3: A multilayer PCB Stackup